The Advances publishes reviews and critical articles covering the entire field of normal anatomy (cytology, histology, cyto- and histochemistry, electron microscopy, macroscopy, experimental morphology and embryology and comparative anatomy). Papers dealing with anthropology and clinical morphology will also be accepted with the aim of encouraging co-operation between anatomy and related disciplines.

Papers, which may be in English, French or German, are normally commissioned, but original papers and communications may be submitted and will be considered so long as they deal with a subject comprehensively and meet the requirements of the "Advances".

For speed of publication and breadth of distribution, this journal appears in single issues which can be purchased separately; 6 issues constitute one volume.

It is a fundamental condition that submitted manuscripts have not been, and will not simultaneously be submitted or published elsewhere. With the acceptance of a manuscript for publication, the publisher acquire full and exclusive copyright for all languages and countries. 25 copies of each paper are supplied free of charge.

Die Ergebnisse dienen der Veröffentlichung zusammenfassender und kritischer Artikel aus dem Gesamtgebiet der normalen Anatomie (Cytologie, Histologie, Cyto- und Histochemie, Elektronenmikroskopie, Makroskopie, experimentelle Morphologie und Embryologie und vergleichende Anatomie). Aufgenommen werden ferner Arbeiten anthropologischen und morphologisch-klinischen Inhalts, mit dem Ziel, die Zusammenarbeit zwischen Anatomie und Nachbardisziplinen zu fördern.

Zur Veröffentlichung gelangen in erster Linie angeforderte Manuskripte, jedoch werden auch eingesandte Arbeiten und Originalmitteilungen berücksichtigt, sofern sie ein Gebiet umfassend abhandeln und den Anforderungen der „Ergebnisse" genügen. Die Veröffentlichungen erfolgen in englischer, deutscher und französischer Sprache.

Die Arbeiten erscheinen im Interesse einer raschen Veröffentlichung und einer weiten Verbreitung als einzeln berechnete Hefte; je 6 Hefte bilden einen Band.

Grundsätzlich dürfen nur Arbeiten eingesandt werden, die nicht gleichzeitig an anderer Stelle zur Veröffentlichung eingereicht oder bereits veröffentlicht worden sind. Der Autor verpflichtet sich, seinen Beitrag auch nachträglich nicht an anderer Stelle zu publizieren. Die Mitarbeiter erhalten von ihren Arbeiten zusammen 25 Freiexemplare.

Les résultats publient des sommaires et des articles critiques concernant l'ensemble du domaine de l'anatomie normale (cytologie, histologie, cyto- et histochimie, microscopie électronique, macroscopie, morphologie expérimentale, embryologie et anatomie comparée). Seront publiés en outre les articles traitant de l'anthropologie et de la morphologie clinique, en vue d'encourager la collaboration entre l'anatomie et les disciplines voisines.

Seront publiés en priorité les articles expressément demandés, nous tiendrons toutefois compte des articles qui nous seront envoyés dans la mesure où ils traitent d'un sejet dans son ensemble et correspondent aux standards des «Revues». Les publications seront faites en langues anglaise, allemande et française.

Dans l'intérêt d'une publication rapide et d'une large diffusion les travaux publiés paraitront dans des cahiers individuels, diffusés séparément: 6 cahiers forment un volume.

En principe, seuls les manuscrits qui n'ont encore été publiés ni dans le pays d'origine ni à l'éntranger peuvent nous être soumis. L'auteur s'engage en outre à ne pas les publier ailleurs ultérieurement. Les auteurs recevront 25 exemplaires gratuits de leur publication.

Manuscripts should be addressed to/Manuskripte sind zu senden an/Envoyer les manuscrits à:

Prof. Dr. A. BRODAL, Universitetet i Oslo, Anatomisk Institutt, Karl Johans Gate 47 (Domus Media), Oslo 1/Norwegen

Prof. W. HILD, Department of Anatomy, Medical Branch, The University of Texas, Galveston, Texas 77550/USA

Prof. Dr. J. van LIMBORGH, Universiteit van Amsterdam, Anatomisch-Embryologisch Laboratorium, Mauritskade 61, Amsterdam-O/Holland

Prof. Dr. R. ORTMANN, Anatomisches Institut der Universität, Lindenburg, D-5000 Köln-Lindenthal

Prof. Dr. T. H. SCHIEBLER, Anatomisches Institut der Universität, Koellikerstraße 6, D-8700 Würzburg

Prof. Dr. G. TÖNDURY, Direktion der Anatomie, Gloriastraße 19, CH-8006 Zürich/Schweiz

Prof. Dr. E. WOLFF, Collège de France, Laboratoire d'Embryologie Expérimentale, 49 Avenue de la belle Gabrielle, Nogent-sur-Marne 94/Frankreich

Advances in Anatomy, Embryology and Cell Biology
Ergebnisse der Anatomie und Entwicklungsgeschichte
Revues d'anatomie et de morphologie expérimentale

Vol. 53 · Fasc. 5

Hans-Werner Denker

Implantation

The Role of Proteinases, and Blockage of Implantation
by Proteinase Inhibitors

With 35 Figures

Springer-Verlag Berlin Heidelberg New York 1977

Priv.-Doz. Dr. med. Dr. rer. nat. H.-W. Denker, Abteilung Anatomie der Medizinischen Fakultät an der RWTH Aachen, Melatener Str. 211, D–5100 Aachen, Federal Republic of Germany

To my Mother

Habilitationsschrift, translated from the German and published with the permission of the Medizinische Fakultät an der Rhein.-Westf. Techn. Hochschule Aachen.
German Title: Mechanismen der Implantation des Säugetierembryos und ihre experimentelle Beeinflussung.

ISBN-13: 978-3-540-08479-2   e-ISBN-13: 978-3-642-66781-7
DOI: 10.1007/ 978-3-642-66781-7

Library of Congress Cataloging in Publication Data. Denker, Hans-Werner, 1941-   Implantation. (Advances in anatomy, embryology, and cell biology; 53/5) Bibliography: p.   Includes index. 1. Ovum implantation. 2. Proteinase. 3. Enzyme inhibitors. 4. Rabbits-Physiology. I. Title. II. Series. [DNLM: 1. Nidation. 2. Peptide hydrolases. 3. Peptide hydrolases--Antagonists and inhibitors. Wl AD433K v. 53 fasc. 5/QS645 D396m] QL801.E67 vol. 53/5 [PQ275] 574.4'08s [599'.03'3] 77-15046

Composition: H. Stürtz AG, Universitätsdruckerei, Würzburg
2121/3321-543210

Contents

Abbreviations

AMCHA	trans-4-(aminomethyl)-cyclohexane-carbonic acid (Ugurol[®])
BANA	benzoyl-arginine-β-naphthylamide
BAPA	benzoyl-arginine-p-nitroanilide
DFP	diisopropyl fluorophosphate
d p.c.	days post coitum
DSI	proteinase inhibitor from dog submandibular glands (dog submandibular inhibitor)
EACA	ϵ-aminocaproic acid
EDTA	ethylene diamine tetraacetate
FBB	Fast blue B
GPNA	glutaryl-phenylalanine-β-naphthylamide
h p.c.	hours post coitum
ImU	international milliunit (−s)
IUD	intrauterine device
KIE	kallikrein-inhibition-unit (−s)
LAP	leucine aminopeptidase
LeuNA	leucine-β-naphthylamide
MS	mucosubstance
NA	β-naphthylamide
NBD	5-nitro-3H-1,2-benzoxathiole-2,2-dioxyde (2-hydroxy-2-nitro-α-toluenesulfonic acid sultone, chymotrypsin titration reagent)
NPGB	p-nitrophenyl-p'-guanidinobenzoate (trypsin titration reagent)
PAA	polyacrylamide
PAS	periodic acid-Schiff reaction
PSTI	bovine pancreatic secretory trypsin inhibitor (Kazal)
SBTI	soybean trypsin inhibitor (Kunitz)
SSPI	boar seminal plasma trypsin-acrosin inhibitor
TCA	trichloroacetic acid
TLCK	p-tosyl-L-lysine-chloromethylketone
TPCK	p-tosyl-L-phenylalanine-chloromethylketone
v/v	volume per volume
w/v	weight per volume

1. Introduction

The establishment of the morphologically and physiologically intimate contact between two genetically different individuals, mother and embryo, which takes place during implantation, has always exerted a fascination on researchers in biology and medicine. Recent years have also seen the beginnings of a more practice-oriented medical interest in this event, as certain methods of contraception whose use is ever increasing, namely intra-uterine devices (IUDs) and post-coital oral contraceptives, are based on it. On one hand, we have fairly well substantiated experimental and clinical evidence that the efficacy of these contraceptives lies with other factors as well, for example their influence on sperm migration and capacitation or their influence on the transportation of the unfertilized egg. On the other hand there appears to be more importance in their influence on the blastocyst and on the early stages of implantation (Carol et al., 1973; Duncan and Wheeler, 1975; Hafez and Evans, 1973; Oettel, 1975). These stages in development are characterized by complex interactions between embryo and mother, which have only lately been more exhaustively investigated and which are still subjects of intense research (Beier, 1973, 1974a; Blandau, 1971a; Finn and Porter, 1975; Hafez and Evans, 1973; Steven, 1975). It is expected that insight into the mechanism of the action of postcoital contraceptives and a possible basis for the development of new concepts in contraception can be gained here.

Historically we owe it to Graf von Spee (1883, 1901) for exciting interest over the question which role the early embryo and uterus play during the contact process (Krüger, 1969; Heinricius, 1914). On his model of the implantation in guinea pigs, which, like that of man, is of the interstitial type, he showed that the trophoblast may play an active and cytolytic role in attachment and invasion. In the guinea pig, pseudopodia-like protrusions of the trophoblast perforate the zona pellucida, make contact with the uterine epithelium, penetrate it and make possible an invasion of the blastocyst in the endometrial stroma where the contact with the maternal blood vessels will be established. The apparent cytolytic activity of the growing trophoblast was emphasized by Spee: the embryo is surrounded in the stroma of the endometrium by an "implantation cavity" on the border of which the uterine tissue shows signs of degeneration and lysis. Spee apparently believed in an indirect influence by the embryo: he spoke of a "biochemical process which is stimulated by the egg" (1901 p. 144). On the other hand he also described reactive processes of proliferation in the endometrium which he found in the region of the implantation zone.

Spee's model gave the discussion about the mechanism of implantation stronger momentum than the observations of others made at this time, regarding the attachment of embryos in various animals with central or eccentric implantation where a disappearance of the uterine epithelium had also been observed (see Hubrecht, 1888, 1889/90, 1909). The physiological discussion begun at this time of the respective roles played by mother and embryo during implantation has not yet reached an end. At the turn of the century the possibility of an involvement of *proteolytic enzymes* of the trophoblast (or endometrium) in man was already under discussion. The first indications for the existence of uterine *proteinase inhibitors* which could possibly regulate enzyme activity were found (Polano, 1907; Grafenberg, 1910; Halban and Frankl, 1910; Frankl and Aschner, 1911; Caffier, 1929a and b). More recently research along these lines was resumed by Schmidt-Matthiesen (1967, 1968, 1970) using human endometrium and placental tissues. In the guinea pig, the pseudopod-like protrusions of the trophoblast which penetrate the zona pellucida were described anew by Blandau (1949, 1971b; Blandau and Rumery, 1957; see also Owers, 1970, 1971, Owers and Blandau, 1971). He found a remarkably high proteolytic activity in the attaching blastocyst. Systematic experimentation, especially with consideration of the other enzymes which could possibly be involved, and a study of the chemistry of the presumable physiological substrate of these enzymes (zona pellucida or its equivalent; other extracellular substances) had yet to be carried out.

In recent years investigation of the implantation process and its biochemical aspects has been intensified. Substrate histochemical investigations, especially in the rabbit, have lead us to imagine that the mucosubstances (glycoproteins) of the *blastocyst coverings* and of the *cell surface* (glycocalyx) function as a barrier or perhaps also as mediators of adhesion (compare discussion, Denker 1970a and b, 1971a–c, 1973, 1975). Böving (1954, 1962, 1963, 1970) developed the hypothesis that in this species the superficial layer of the blastocyst coverings (so-called "gloiolemma") changes its viscosity due to a local rise in pH during the beginning of implantation, becomes "stickier" and is responsible for the first loose adhesion of the blastocyst on the uterine epithelium . A change in the histochemical characteristics of the blastocyst coverings during implantation is actually observed. The hypothesis has been put forward that glycosidases and proteases, which show considerable activity in the endometrium and blastocyst, play a role here and operate synergistically (Denker, 1970a and b, 1971a–c, 1973). Particular attention was paid to the notably high *proteinase activity of the attaching blastocyst* in the rabbit (Denker, 1969, 1970b, 1971c, 1972, 1974b, 1975, 1976a; Denker and Hafez, 1975). Other authors proposed a key role for the proteinases of the uterine secretion (Mintz, 1971, 1972; Kirchner, 1972a, 1975; Kirchner et al., 1971). An experimental investigation of whether or not these proteinases or other enzymes actually have such an important role in the process of implantation was, however, lacking.

This question is the central one of the following paper.

The experiments were carried out principally in the rabbit, as in this species extensive investigation of the morphology, histochemistry and biochemistry of implantation had been undertaken previously. In addition the rabbit is a reflex ovulator: ovulation takes place nearly exactly 10 hours post coitum (h. p. c.), so that it is possible to determine the stage of pregnancy fairy accurately. Further, at approximately 5 mm in diameter, the rabbit blastocyst is relatively large compared to other laboratory animals, which makes experimentation easier. For comparison, a limited number of experiments were carried out on embryos of the *cat,* who is likewise a reflex ovulator (ovulation 25–28 hrs p.c.).

2. Materials and Methods

2.1. Laboratory Animals and Experimental Outline

All animals were kept in single cages at controlled temperature with a constant light-dark rhythmn (rabbits: 12:12 hours, cats: light phase 14 hrs., dark phase 10 hrs.) and were fed with standardized pellet food.

Rabbits: Sexually mature females were mated to two fertile males each. In order to induce pseudopregnancy, some does were mated to vasectomized males. At a particular time post coitum (days post coitum = d p.c.) the animals were killed by exsanguination. The uterus was then immediately removed.

9

Uteri and blastocysts from normal pregnancy were taken from 30 crossbred females.

A total of 69 sexually mature females were treated in vivo with proteinase inhibitors. First the appropriate type of application (injection in the uterine lumen or "slow-release IUD") and application timing for the inhibitors was experimentally determined for 10 females of different races and crosses. The "slow-release devices" used in six of the animals were beads with a diameter approximately that of a blastocyst in the stage seven days p.c., i.e. about 4.5 mm. A 5 % aqueous agarose solution was prepared, autoclaved and in certain cases mixed with the desired amount of inhibitor. The beads were formed in a mold under sterile conditions. Identical beads without inhibitor served as controls. The IUDs were used immediately after manufacture. It is a known fact that slow-release devices of various types have a rate of pharmacon release which is rapid at first and then settles into a lower, constant rate. In our preliminary experiments animals were sacrificed 24 hours after insertion of the IUDs. As we discovered that even at a concentration of 10 mg/ml (Trasylol®) in the agar beads and upon the application of 2–3 beads per uterus (time of application: 6–6 1/2 d p.c.) not all blastocysts from one uterus showed an impaired implantation, but only those which were adjacent to IUDs, we gave the "slow-release device" method up in favor of an injection of the inhibitor solution into the uterine cavity.

The following plan was therefore drawn up for the central experiment (compare Fig. 1, Tab. 1): a total of 59 pure-bred Alaska females (our own strain), 6–10 months old, weighing 2.4–4.2 kg were bred as outlined above. Exactly 6 days and 12 hours p.c. (i.e. half a day before implantation) the animals are laparotomized after neuroleptic premedication (Decentan®) and administration of a thiobarbiturate anaesthetic (Thiogenal®) (see Zimmermann, 1964; Gottschewski and Zimmermann, 1970). For a list of the proteinase inhibitors which were used and their dosage, see Tab. 1 and 11. With the exception of NPGB all inhibitors were dissolved in sterile 0.9% NaCl. NPGB was first dissolved at 10-fold concentration in N,N'-dimethylformamide and then diluted with NaCl solution to 1/10 of this concentration. 0.6 ml of inhibitor solution prepared in this fashion was injected into the uterine lumen, using three injections of 0.2 ml each on the tubal and vaginal ends and in the middle of the uterus. For this purpose the side with the greatest number of corpora lutea was usually chosen. When uterine flushings were to be harvested, the side with the least number of corpora lutea was chosen. Into the other uterus, as a control, 3 x 0.2 ml of 0.9% NaCl (in the case of NPGB 10% N,N'-dimethylformamide in NaCl solution, v/v) was injected. Exactly 7 1/2, 8 1/2, 9 1/2 or 11 1/2 d p.c., the animals were sacrificed to obtain the uteri and blastocysts.

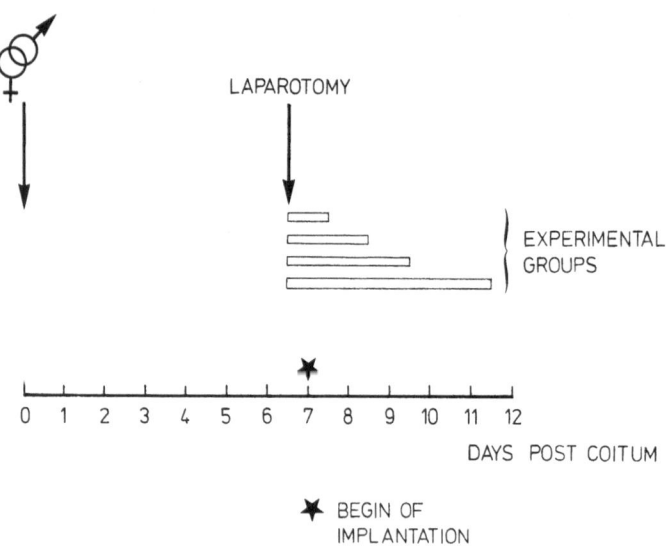

Fig. 1. Outline of experiments on the inhibition of blastocyst proteinase and of implantation in the rabbit through the application of proteinase inhibitors in vivo

Table 1. Number of animals and outline of in vivo experiments with proteinase inhibitors in the rabbit. Application of the inhibitors 6 d 12 h p.c. through intrauterine injection. For dosage see Tab. 11 and 12

A. Animals from which material was taken for histological, histochemical and electron microscopic investigations

Inhibitor	7d12hr p.c.	8d12hr p.c.	9d12hr p.c.	11d12hr p.c.	total
Trasylol®	7	5	2	2	16
antipain	5	5	3	2	15
NPGB	3	1	1		5
SSPI	1				1
EACA	3				3

B. Animals from which uterine flushings and unattached blastocysts were harvested

Inhibitor	6d12hr p.c. (0)[a]	6d13hr p.c. (1)[a]	6d20hr p.c. (8)[a]	7d12hr p.c. (24)[a]	8d12hr p.c. (48)[a]	9d12hr p.c. (72)[a]	total
Trasylol®	3	1	1	4	3	1	13
antipain	3	0	4	4	3	0	14

[a]Time in hours after application

From 8 animals uterine flushings as well as material for histological and histochemical purposes were taken. Therefore the sum of A and B is larger than the total number of animals

Cats: Material from a total of seven pregnant cats in the stages 12–14 d p.c. was studied (12 d p.c. 2 animals, 13 d p.c. 3 animals, 14 d p.c. 2 animals). Uteri from 4 other cats who were similarly bred but who 12–14 d p.c. contained no blastocysts were used for comparison. The animals were bred during oestrus, determined by daily control of vaginal smears, with tame, trained tom cats (for further details see Sojka et al., 1970; Hamner et al., 1970; Khera, 1973). The animals were killed with an overdose of sodium pentobarbital and the uteri were removed at once.

All cat uteri were immediately frozen with liquid nitrogen to be cut on a cryostat for morphological and substrate histochemical as well as for enzyme histochemical investigations, while with rabbit material different pretreatments were used for different types of studies.

2.2. Morphological Investigations

For observation under the light microscope, pieces of rabbit uteri were fixed by immersion in formol-calcium acetate (1 % w/v) at 4°C for 48 hrs, paraffin embedded and cut 7 μ thick (compare Denker, 1970a). Cryostat sections taken from cat uteri, on contrast, were post-fixed with 10 % aqueous formaldehyde. Stains used include haemalaun-eosin, azan, Masson-Goldner (compare Romeis, 1968).

For morphological investigations using semi-thin and thin sections, rabbits were deeply anesthetized with sodiumpentobarbital and perfused through the aorta thoracica with 2.5 % glutaraldehyde in 0.1 M cacodylate buffer at pH 7.3−7.4 containing 0.05 % $CaCl_2$ at 0°C. In many cases the blastocyst cavity was directly perfused with the same fixative by puncturing it with 2 hypodermic needles. The tissues were then rinsed in cacodylate buffer, fixed in 2 % osmium tetroxide, dehydrated in ethanol and propylene oxide, embedded in Araldite, and semi-thin and thin sections were cut on the Reichert ultramicrotome 0m U3. 0.6 μ semithin sections were stained with azure II-methylene blue (Richardson et al., 1960) or with toluidine blue. Ultrathin sections were contrasted with uranyl acetate and lead citrate and viewed with a Philips electron microscope 300 at 80 KV and with a Zeiss EM 10A at 60 KV.

2.3. Histochemical Investigations

Substrate histochemical investigations on the distribution and characteristics of *mucosubstances* (MS) were carried out on paraffin sections (see above) of pieces of rabbit uteri with blastocysts. Post-fixed cryostat sections were used, however, in the cat (see 2.2.). For the details of the histochemical reactions compare Denker (1970a).

For enzyme histochemistry pieces of uteri were frozen quickly in liquid nitrogen and sectioned 14 μ thick on the cryostat.

The gelatin substrate film test (Denker, 1969, 1974a) served as the histochemical *proteinase* assay, with the following modifications: 0.2 ml gelatin per slide was used in the experiments; washing (step A5) was done in deionized water rather than tap water; the incubation time was 105 minutes at 38°C; the reaction was stopped in saturated aqueous $HgCl_2$ solution at 4°C; the stain was 0.5 % toluidine blue in a mixture of equal parts of absolute ethanol and 0.2 M borate buffer pH 10.0. For investigating inhibition caused by various inhibitors in vitro, the substrate films were impregnated with a solution containing the inhibitor and dried again before mounting (Denker, 1976a). For investigating the pH dependence of the enzymatic reaction the gelatin films were similarly soaked in 0.1 M phosphate buffer at various pH's. *Amino acid arylamidases* (EC classification not yet final; probably related to EC 3.4.11.2, see 4.4.1.) were studied using unfixed sections or after section freeze substitution; incubation was done in liquid incubation medium, or a membrane method was applied. Substrates: β-napthylamides of various amino acids (alanine, arginine, aspartic acid, glutamic acid, glycin, leucine, lysine, phenylalanine, tryptophane, tyrosine; in the case of leucine the faster coupling leucine-4-methoxy-β-napthylamide was used in addition to the simple β-napthylamide), "simultaneous coupling" with Fast Blue B (FBB) or Fast Garnet GBC (Denker and Stangl, 1976).

Glycosidases: Investigation done with formol-calcium-fixed cryostat sections with the exception of α-amylase, which was investigated on unfixed sections. For materials, methodological details and references see Denker (1971a, b) and Pearse (1972).

2.4. Biochemical Investigations

Uterine secretion, endometrium and blastocysts were harvested separately for biochemical tests. To obtain *uterine secretion*, a glass tube was inserted from one end of the uterus into the lumen, which was then *flushed* with 5 ml of 0.9 % NaCl solution from the other end. In this way was it possible to avoid contamination of the flushing fluid with blood or detritus from the cut end (see Beier, 1968a). Flushing of the entire uterus was not done at the post-implantation stages in order to avoid a rupture of attached blastocysts. At these stages (i.e. normally from 7 d p.c. on, or from 8 1/2 d p.c. on when treated with inhibitors) only the uterine sections lying between implantation sites were flushed with 1−3 ml (depending on the length of the uterine section). The uterine secretion obtained was frozen in dry ice after removal of any blastocysts present. For enzymatic tests the secretion was centrifuged at 3,000 g and 4°C for 10 minutes, and the supernatant was used. *Endometrium* was removed from the opened uterus with a pair of curved scissors. Cells were disrupted by freezing with dry ice and thawing and by homogenizing in an ice-cooled Teflon-glass Potter homogenizer and centrifuged as above (see van Hoorn and Denker, 1975). After being flushed out of the uterus the *blastocysts* were washed in 0.9 % NaCl solution, then put into fresh saline solution and opened with fine forceps; the blastocyst coverings were teased apart from the blastocyst tissue. The latter was often additionally separated into embryonic disc and trophoblast (with extraembryonic entoderm). Whereas contamination here of the embryonic disc material with trophoblast cannot be avoided with certainty, contamination of the trophoblast with embryonic disc cells can be. Homogenization was done with an electric stirrer (Eppendorf system) as well as by additional freezing and thawing.

All steps in the procedure were carried out at $0-4°C$, and precooled solutions were always used. The material was quickly frozen in liquid nitrogen or dry ice for storage at $-30°C$.

Quantitative Enzyme Tests

Amino acid arylamidase activity was measured with leucine-β-naphthylamide (LeuNA) as substrate in veronal acetate Michaelis buffer pH 7.0 (determination of the β-naphthylamine released with the Bratton-Marshall reaction (see van Hoorn and Denker, 1975)).

Endopeptidase activity was determined against casein with the azocasein test (Fritz et al., 1974b). One proteinase milliunit (mU^{ACas}) was defined as the amount of enzyme which, under the given conditions (volume 0.4 ml, after the addition of trichloroacetic acid 1.0 ml; 10 min. incubation at $37°C$) releases an amount of acid-soluble peptides that cause a rise in the extinction at 366 nm of 0.001 per minute.

Determination of Proteinase Inhibitor Activity

The activity of proteinase inhibitors in the uterine secretion was determined using bovine trypsin and benzoyl-arginine-p-nitroanilide (BAPA) as substrate (Fritz et al., 1974b). Care had to be taken in the determination of antipain because the inhibitor constant is fairly high (as opposed to Trasylol® and SSPI), so that displacement phenomena had to be avoided in the experiment. Inhibitor activities were always read from the practically linear beginning of the standard graph, and the conditions of the reaction under which the values were obtained were rigidly standardized.

All experiments were carried out at least three times and the values averaged. The activity of the inhibitor was based on the volume unit of the uterine secretion on the one hand, and on its content of proteins which could be precipitated by TCA (specific activity) on the other hand. The TCA precipitation was necessary as the inhibitors used in this experiment are peptides themselves but are TCA-soluble.

Determination of the *protein concentration* was done according to Lowry (see Zak and Cohen, 1961), using bovine serum albumin as a standard.

Electrophoresis

Agar gel electrophoresis was carried out according to the standard immunoelectrophoresis procedure (1.5 % agarose, 0.4 M veronal-acetate-buffer pH 8.2; 0.1 % gelatin was added to the agarose for the proteinase test. After separation over 2 hours at $0-4°$ C the sample was incubated for $18-24$ hours at $38°C$ in a sealed wet chamber. Staining was done with amido black.

Polyacrylamide (PAA)-gel-disc-electrophoresis in micro scale was carried out in 5 μl Drummond microcaps (separation gel: 20 % PAA, pH 8.8; spacer gel: 5 % PAA, pH 6.7) according to Neuhoff (1968, see Petzoldt, 1974, Petzoldt et al., 1972). *The gelatin substrate film test* (Denker, 1974a) was preferably used for localization of proteinase fractions in the gel. In order to obtain a staining of protein bands as well as a proteinase reaction from each individual gel, gels were sliced longitudinally on a cryostat (at a thickness of 14 μ). Another series of gels was used unsectioned to visualize fractions which hydrolyze synthetic amide substrates (benzoyl-arginine-β-naphthylamide = BANA, or glutaryl-phenylalanine-β-naphthylamide = GPNA; diazonium salts: Fast Garnet GBC or Fast Blue B) (for details of the procedures see Denker and Petzoldt, 1977).

PAA-gel-disc-electrophoresis in macro scale was carried out according to Maurer (1971) (spacer gel 2.5 % PAA, pH 6.7; separation gel 7.5 % PAA, pH 8.9; separation time 30 min.; protein staining with amido black). 0.1 % gelatin was polymerized into the gel for the localization of proteinases. In this case the separation was carried out at about $4°C$; the gels were finally incubated in 0.1 M phosphate buffer pH 7.0 for 1–2 hours and then stained in a mixture of equal parts of a 0.2 % light-green solution in 0.5 % sulfosalicylic acid, of a 0.2 % Ponceau-red solution in 5 % trichloroacetic acid, and of a 0.1 % amido black solution in 5 % trichloroacetic acid (after a suggestion made by Stegemann, 1968; Klockow, personal communication; Fritz et al., 1975b).

2.5. Chemicals

Enzymes: Amino acid arylamidase from porcine kidneys (α-aminoacyl-peptide hydrolase, microsomal, EC 3.4.11.2) (Boehringer/Mannheim No. 15075, ca. 4 U/mg using L-leucine-p-nitroanilide as substrate); carboxypeptidase A from bovine pancreas (EC 3.4.12.2) (Boehringer/Mannheim No. 15441, ca. 35 U/mg using hippuryl-L-phenylalanine as substrate, traces of chymotrypsin and trypsin not detectable); carboxypeptidase B from porcine pancreas (EC 3.4.12.3) (Boehringer/Mannheim No. 15265, ca. 150 U/mg using hippuryl-L-arginine as substrate, contaminating activities: carboxypeptidase A < 2 %, after treatment with DFP chymotrypsin and trypsin not detectable); α-chymotrypsin (EC 3.4.21.1), bovine (Serva, Heidelberg, No. 17160, ca. 45 BTEE-U/mg, ca. 600 Anson-U/mg); elastase from pancreas (EC 3.4.21.11) (Merck No. 24552, 80 U/mg using orcein-elastin as substrate); collagenase (EC 3.4.24.3) (Calbiochem No. 234136, 83 U/mg using collagen as substrate, amount of nonspecific proteases 0.01 %); leucine aminopeptidase from porcine kidneys (α-aminoacyl-peptide hydrolase, cytosol, EC 3.4.11.1) (Sigma No. L-1503, type V, 123 U/mg using L-leucine amide as a substrate, free of BAEE-, BTEE- or TAME-splitting activity); papain from papaya carica (EC 3.4.22.2) (Boehringer/Mannheim No. 15464, ca. 30 U/mg using BAEE as a substrate; pepsin (EC 3.4.23.1), porcine (Boehringer/Mannheim No. 15445, ca. 2500 Anson-U/mg); plasmin (EC 3.4.21.7), human (AB KABI, Stockholm, ca. 15 casein units/mg protein); trypsin (EC 3.4.21.4), bovine (Serva, Heidelberg, No. 37260, free of chymotrypsin, 180 EU/mg, ca. 2500 Anson-U/mg).

Enzyme substrates: Azocasein (Serva No. 14390); N-α-benzoyl-L-arginine-β-naphthylamide · HCl p.A. (L-BANA) (Serva No. 14622); N-α-benzoyl-DL-arginine-β-naphthylamide · HCl p.A. (DL-BANA) (Serva No. 14630); N-α-benzoyl-L-arginine-4-nitroanilide-hydrochloride (BAPA) (Merck No. 10754); casein according to Hammarsten (E. Merck, Darmstadt, No. 2242); gelatin (E. Merck, Darmstadt, No. 4070); glutaryl-L-phenylalanine-β-naphthylamide (GPNA) (E. Merck, Darmstadt, No. 4245 or K & K Laboratories/ICN Pharmaceuticals No. 24820); L-leucine-4-methoxy-β-naphthylamide · HCl (Serva No. 27727); L-leucine-β-naphthylamide · HCl p.A. (Serva No. 27730).

Inhibitors: ε-Aminocaproic acid (EACA) (Serva Nr. 12548); trans-4-(aminomethyl)-cyclohexane carbonic acid (trans-AMCHA) (Ugurol®, Bayer AG); diisopropyl fluorophosphate (DFP) (C. Roth, Karlsruhe, No. 1–5465); 5-nitro-3H-1, 2-benzoxathiole-2, 2-dioxide (NBD) (2-hydroxy-5-nitro-α-toluenesulfonic acid sultone) (Eastman Kodak Co. No. 10335); p-nitrophenyl-p'-guanidinobenzoate (NPGB) (Merck No. 10562); ovomucoid from chicken (Worthington No. 0I 35D952); tosyllysine chloromethyl ketone (TLCK)= L-1-chlor-3-p-tosylamido-7-amino-2-heptanone (Serva No. 17013); tosylphenylalanine chloromethyl ketone (TPCK)= L-1-chlor-3-p-tosylamido-4-phenyl-2-butanone (Serva No. 17016); trypsin inhibitor from soybeans (Kunitz, SBTI) (Merck No. 24020); bovine trypsin-kallikrein-inhibitor (Trasylol®, Bayer AG, charge SMU 224/9, 6900 KIE/mg = 3860 ImU/mg); antipain: a filtrate of actinomycetales cultures donated by Dr. E. Truscheit and Dr. W. Wingender, Bayer AG, Elberfeld, Art.-No. BAY f 7412, contains about 65 % antipain, 670 FIP-trypsin-inhibitor-units/mg = 21400 KIE/mg = 12000 ImU/mg; α_1-antitrypsin and α_1-antichymotrypsin were donated by Prof. Dr. H. G. Schwick and Dr. N. Heimburger (Behring-Werke, Marburg) (O.P. No. 18371 and 1171).

Prof. Dr. H. Fritz, Munich, donated the trypsin-acrosin inhibitor from porcine seminal plasma (SSPI, a mixture of fractions I and II as well as pure fraction II), the secretory trypsin inhibitor from bovine pancreas (PSTI, Kazal), the dog submandibular gland inhibitor (DSI) and the microbial inhibitors leupeptin, chymostatin and pepstatin (samples from H. Umezawa, Tokyo).

Miscellaneous: Albumin from bovine serum, crystal., pure 99.4 % (Serva No. 11920); Fast Blue B salt · (BF$_4$)$_2$, pure, free of heavy metals (Serva No. 21269); Fast Garnet GBC salt, pure (Serva No. 21290). The remaining chemicals not listed separately here were purchased in the purest available forms currently on the market (in general analytic grade).

3. Results

3.1. Morphology of Normal Implantation

3.1.1. Rabbit

The rabbit has a uterus duplex with two separate cervices. The *endometrium* forms a series of longitudinal ridges or folds with cross furrows. Seen somewhat ideally one can recognize 3 pairs of longitudinal ridges: to either side of the axis of symmetry which runs from the mesometrium to the opposite (antimesometrial) side lies a placental fold, an obplacental fold and a paraplacental fold (Duval, 1889a; Minot, 1889; Doorman, 1893; Assheton, 1895; Klein, 1933; Denker, 1970a; Beier, 1973). The placental fold is the most developed even in the non-pregnant animal. As scanning electron microscopical studies impressively show, the longitudinal ridges are divided up by cross furrowing (Busch et al., 1977). This is particularly obvious in non-pregnant animals, which results in our not always seeing the typical pattern of 2 x 3 ridges in cross sections of such uteri (Denker, 1970a).

In *pregnancy,* before implantation of the embryos the endometrium undergoes an impressive transformation, which is at first regulated by the stage-specific changing maternal levels of estrogens and gestagens, while the embryo itself exerts little definitive effect on the surrounding maternal tissues (Beier, 1970a, 1971; Beier et al., 1971, 1972b; Kühnel et al., 1971; Davies and Hoffman, 1973, 1975): The differences between normally pregnant uteri (with embryos) and pseudopregnant uteri (without embryos) are minimal at this point and concern mainly the speed with which the individual transformation and secretion stages follow one another. The morphological characteristic of transformation is a filigree-like change in the originally massive, stroma-rich endometrial folds, which leads to an important enlargement of the surface covered with epithelium (Duval, 1889a; Fraenkel, 1903; Denker, 1970a; Petry et al., 1970). In contrast to the oestrus phase, from about the 5th day p.c. on the cavum epithelial cells show characteristic protrusions of the apical cytoplasm. They have been interpreted as the morphological equivalent of an aprocine extrusion (Beier et al., 1972a; Beier, 1973). In the following days, the cell membranes between adjacent uterine epithelial cells begin to disappear, especially at the site of implantation, so that the cavum epithelium becomes transformed into a syncytium. This process is, however, not yet complete at the beginning of antimesometrial implantation (7 d p.c.) (see below; 4.4.2.2.2.; see Duval, 1889a; Schoenfeld, 1903; Böving, 1962, 1963; Larsen, 1962).

The proliferation of the endometrium is maximal in the region of the placental fold, and is in the beginning uniform throughout the entire length of the uterus suggesting that it is independent of the presence of an embryo. At 7–8 d p.c., as a result of local stimulation whose (chemical?) nature is as yet unknown, a further proliferation occurs at the site of implantation which finally leads to the development of the maternal half of the placenta through further morphological transformation and the appearance of decidual cells (see below).

Implantation in the rabbit occurs in two distinct processes separated clearly in space and time: 1.) the development of a *yolk sac placenta* in the abembryonic-antimesometrial region, and 2.) the development of the definitive *chorioallantoid placenta* mesometrially at the embryonic pole. A haemochorial contact is established in both cases, however the yolk sac placenta is only an ephemeral phenomenon which is later replaced by the chorioallantoid mesometrial placenta. The details of rabbit embryo implantation which can be observed under the light microscope have been exhaustively described in many investigations (Masquelin and Swaen, 1880; van Beneden and Julin, 1884; Duval, 1889a and b, 1890; Minot, 1889; Assheton, 1895; Maximow, 1900; Schoenfeld, 1903; Mossman, 1926; Böving, 1962; Denker, 1970a and b). The rabbit exhibits the *central type of implantation* for which it is characteristic that the blastocysts increase in size and fill out the uterine lumen (Figs. 2 and 3). In the rabbit the walls of the uterus including the myometrium are stretched by the considerable expansion of the blastocysts even before implantation, so that the position of the individual blastocysts is recognizable from the outside 6–6 1/2 d p.c. At this point the still freely movable blastocysts are, through peristalsis-like movements of the myometrium, *evenly distributed* over the length of the uterus (Böving, 1954, 1963). Until 6 2/3 d p.c. they are still surrounded by coverings of extracellular material. These *blastocyst coverings* in the rabbit are not identical with the *zona pellucida.* The latter, which already encompasses the embryo in the ovary, becomes considerably thinner with the expansion of the rabbit blastocyst (from a diameter in the morula of ca. 120 μ to ca. 5 mm at implantation), and it is questionable whether it can still be detected. In this species, however, during passage down the tubes, i. e. still in the cleavage stages, an additional voluminous covering of thickened tubal secretion, the *mucoprotein layer,* is added to the zona pellucida. According to Böving (1957, 1959, 1963) another layer of uterine secretion material is added later on (so-called *gloiolemma*) and is assumed to play a role in the primary adhesion of the blastocyst to the uterine epithelium (see also 4.2.).

Morphology of the Antimesometrial-Abembryonic Implantation (Obplacentation)

7 d p.c. by some unknown mechanism, the embryo orients itself with the embryonic disc mesometrially and with the abembryonic pole across from it, antimesometrially (Fig. 2, 3). Now the *blastocyst coverings* begin to dissolve in the abembryonic-lateral region, at first near the *trophoblastic knobs,* syncytial elements of the trophoblast, then also between the knobs (Böving, 1963; Denker, 1970a and b, 1974b, 1975). The *blastocyst coverings* are composed of several *layers* before lysis which can be told apart by their structure and staining characteristics. 6 1/2 to 7 d p.c. at least two layers are recognizable in paraffin sections, and often also an additional, thinner middle layer (see 3.2.1.). Three layers are always recognizable in semi-thin sections stained with toluidine blue or in thin sections contrasted with uranyl acetate and lead

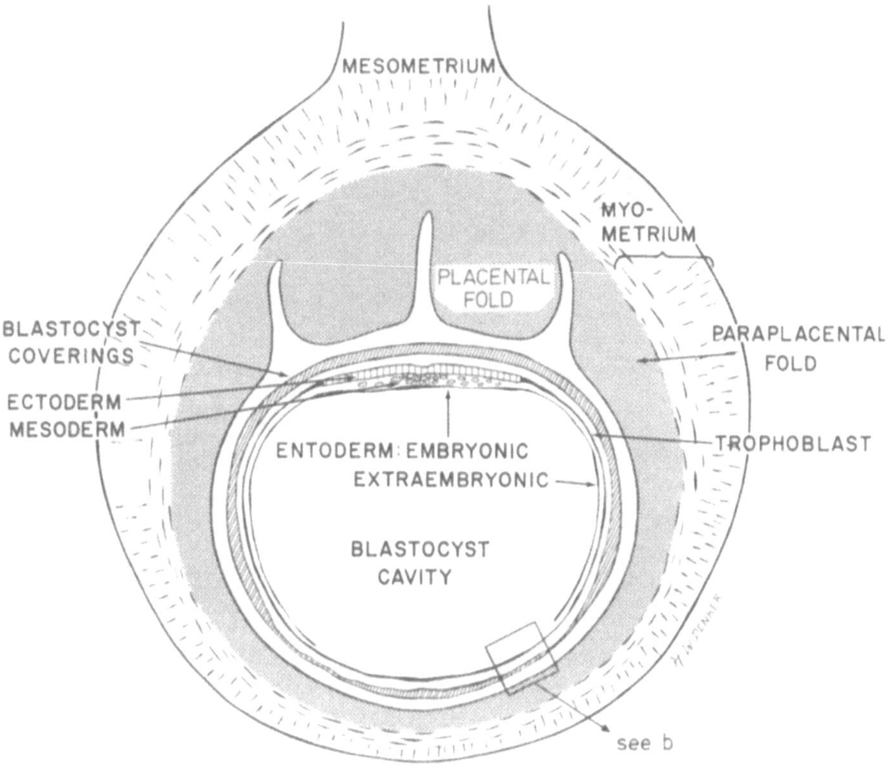

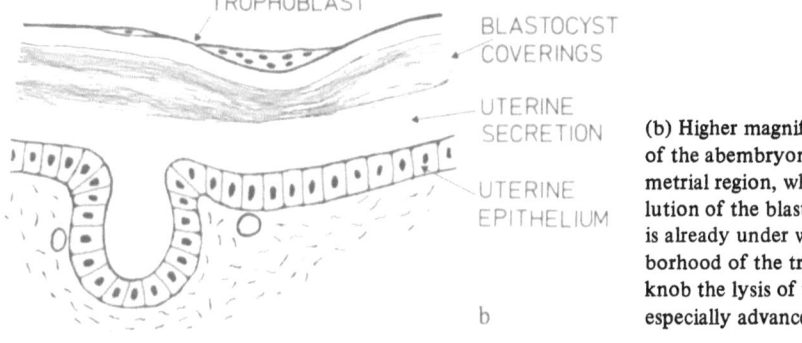

Fig. 2a and b. Schematic drawing of blastocyst and uterus in the rabbit at the beginning of the abembryonic (antimesometrial) phase of implantation 7 d p.c.

(a) Low magnification view of a cross section of an implantation site

(b) Higher magnification of part of the abembryonic-antimesometrial region, where the dissolution of the blastocyst coverings is already under way. In the neighborhood of the trophoblastic knob the lysis of the coverings is especially advanced

citrate. The middle layer shows up in the electron microscope as having a rough structure in which fibrillary elements oriented parallel to the blastocyst surface predominate (Fig. 20).

The first sign of the beginning of dissolution is that the *layering of the blastocyst coverings disappears*, that they swell up and that their inner and outer borders become indistinct (Fig. 5). These unclear-appearing parts of the coverings are at first only found over the trophoblastic knobs. This early stage can be especially well observed when the

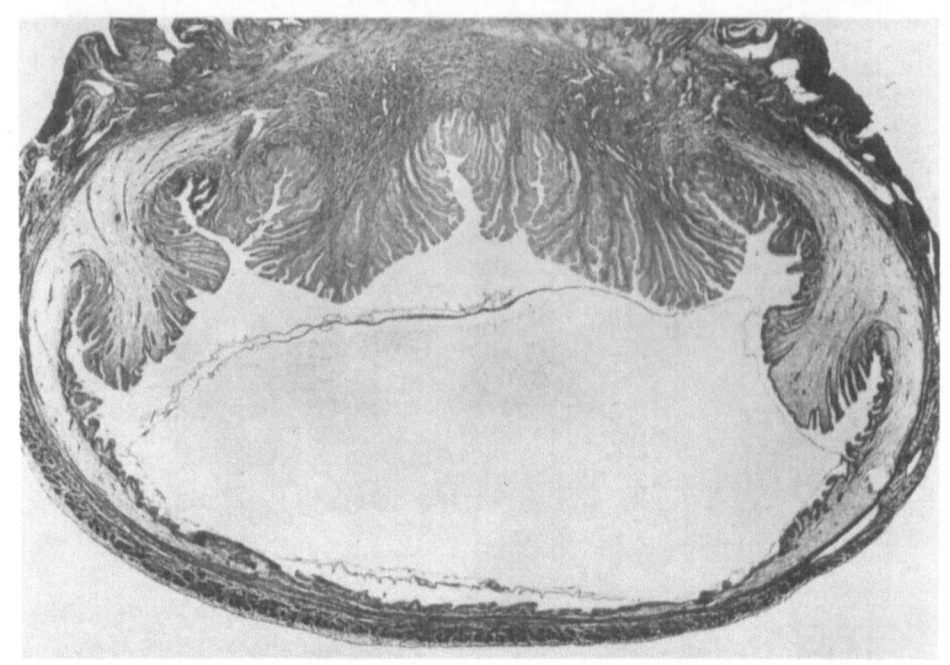

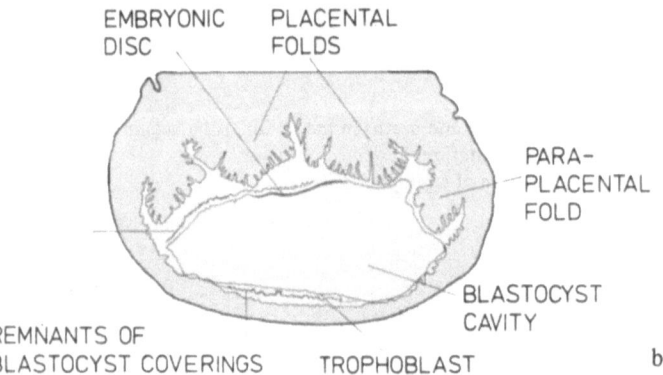

EMBRYONIC PLACENTAL
DISC FOLDS

PARA-
PLACENTAL
FOLD

BLASTOCYST
CAVITY

REMNANTS OF
BLASTOCYST COVERINGS TROPHOBLAST b

Fig. 3a and b. Rabbit, normal control animal, cross section through uterus and blastocyst (implantation site), morphology. 7 1/3 d p.c.
(a) paraffin section, alcian blue 0.2 M MgCl$_2$, X 17; (b) diagram. The strongly expanded blastocyst has widened the uterine lumen and stretched the uterine wall. The well-developed placental folds can easily be recognized in the cross-section of the uterus. They contain especially large amounts of strongly staining mucosubstances in the stroma. The obplacental folds have disappeared. The trophoblast has attached itself to the endometrium in the antimesometrial-lateral regions. There the blastocyst coverings are dissolved whereas above the embryonic disc and the neighboring trophoblastic region (mesometrially) and exactly at the abembryonic pole remnants of the coverings still remain (for a slightly earlier phase see Fig. 2)

dissolution of the coverings is partially inhibited in vivo by minimal doses of proteinase inhibitors (see 3.3.2.). Normally the layering in those parts of the coverings between the trophoblastic knobs also becomes indistinct shortly thereafter, and can then only be found over the embryonic disc up to 7 1/2 d p.c. In the abembryonic hemi-

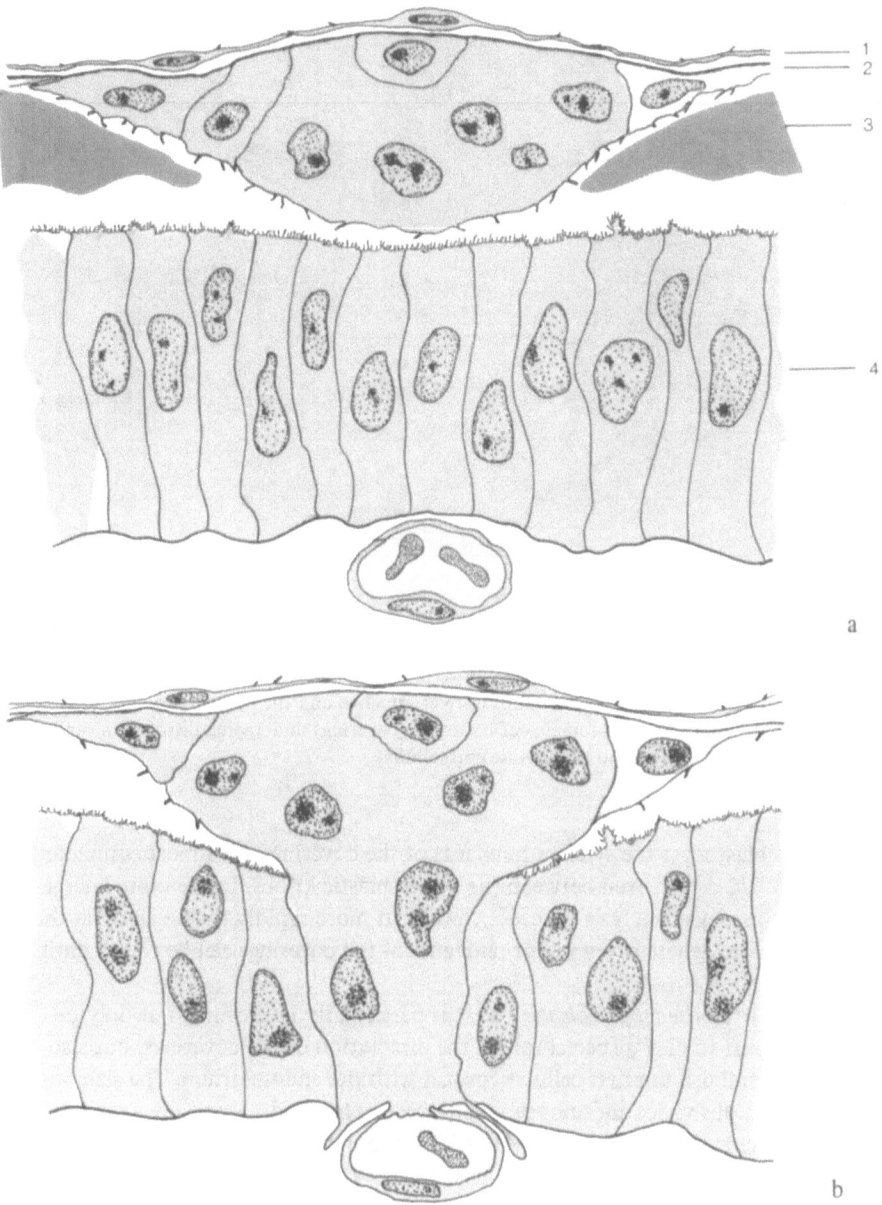

Fig. 4a and b. Slightly schematic drawing of details of the spatial relationship between the abembryonic trophoblast and the antimesometrial endometrium in the rabbit as seen in thin and semithin sections directly before (a) and after (b) attachment, 7–7 1/3 d p.c.

(a) The trophoblastic knob (2) is still partly composed of individual cells, although these have already largely fused. In the direction of the blastocyst cavity (above) a basal lamina and the entoderm lie underneath it; although entoderm cannot be found in all parts of the abembryonic hemisphere (see Fig. 2a). Remnants of the blastocyst coverings (3) are still recognizable to either side of the trophoblastic knob, while the knob itself closely approaches the uterine epithelium which is covered with a lawn of microvilli (4). Note the position relative to a subepithelial capillary. (b) Invasion of a trophoblastic knob into the abembryonic endometrium. The figure follows Fig. 4, but shows a more advanced stage. After the fusion of the trophoblastic knob with the uterine epithelial cell which is located atop the subepithelial capillary shown, the cytoplasm of the trophoblastic knob penetrates the basal lamina and reaches the blood vessel which it erodes

19

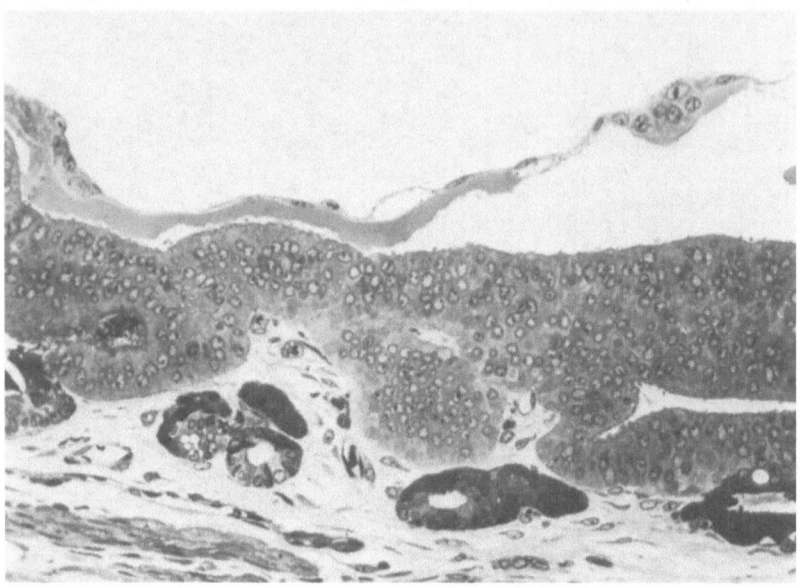

Fig. 5. Morphology of normal implantation in the rabbit (control uterus, see 3.3.2.), 7 1/2 d p.c., semi-thin section, toluidine blue, X 220. Abembryonic trophoblast and the antimesometrial endometrium are shown. Two trophoblastic knobs not yet attached and the cytotrophoblast lying between can be recognized. Swollen remains of blastocyst coverings undergoing dissolution can be seen lying between the trophoblast and the uterine epithelium

sphere of the blastocyst the swollen remnants of the coverings are almost completely dissolved at 7 1/2 d p.c., even between the trophoblastic knobs. In the more lateral parts of the blastocyst the lysis proceeds somewhat more rapidly than exactly at the abembryonic pole, where often minor remnants of the coverings can be found until 7 1/2 d p.c.

The structure of the trophoblastic knobs is particularly interesting, not only because they appear to play a special role in the dissolution of the coverings, but also because they establish the first cellular contact with the endometrium. The size of the trophoblastic knobs varies enormously at all phases. One finds, especially at the beginning of attachment (7 d p.c.), several which are only the size of 2–3 uterine epithelial cells, and there are all transitional sizes between that and those which are larger than 20 epithelial cells. 7 1/2 d p.c. the larger knobs predominate. Trophoblastic knobs come about through cell fusion after a preliminary increase in the volume of the nucleus and cytoplasm of individual trophoblastic cells (see Enders and Schlafke, 1971). Remnants of the cell membrane with desmosomes still mark the spot for a while where the original cell borders were. Not all of the cells in the region around a knob undergo fusion, however, but rather cytotrophoblast cells remain randomly dispersed especially on the edges and at the base of the trophoblastic knob facing the blastocyst cavity (see Fig. 4). But the cytotrophoblast usually forms no continuous covering on the syncytiotrophoblast; and certainly not on the outer surface, as was postulated from light microscopical investigations (see Böving, 1970; Böving and Larsen, 1973), and mostly not on the side facing the blastocyst cavity either. On the borders between individual cells of the trophoblastic knobs, and especially at their

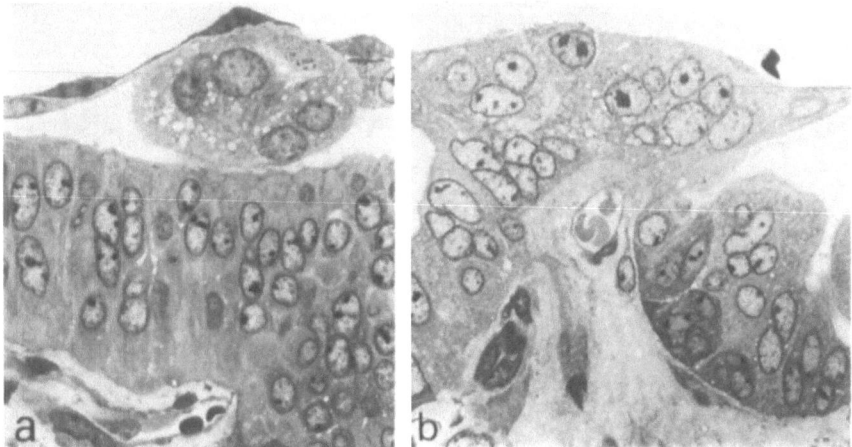

Fig. 6a and b. Rabbit, normal implantation (control uterus, see 3.3.2.), morphology, 7 1/2 d p.c., antimesometrial-abembryonic region, semi-thin section, toluidine blue, X 660.
(a) The blastocyst coverings have disappeared. The uterine cavum epithelium is cylindric and still cellularly differentiated. Underneath is a subepithelial capillary. The trophoblastic knob shown is just touching the uterine epithelium, but it is not yet attached. It is covered with entoderm in the region facing the blastocyst cavity (see Fig. 2). (b) Trophoblastic knob in advanced invasion phase. The uterine epithelium has been overcome, and contact with the subepithelial capillary has almost been established. In the uterine cavum epithelium cell fusion seems to be under way, but is obviously not yet completed. The cytotrophoblast keeps its distance

base, one encounters bizarre, labyrinthine and tortuous infoldings in the surface membrane. In this latter region we find other large, nodular protuberances (blebs) of the surface cytoplasmic areas, which usually contain few organelles. They are round in cross-sections, and their contents appear often scant. We find abundantly all intermediary forms of equally round, membrane enclosed, vesicular structures with flocculous contents of varying electron density (Fig. 7). The organelles contained in trophoblastic knobs are similar to those of the cytotrophoblast between them. Characteristic are the manifold round to long crystalloid inclusions, which have already been described in detail (Enders, 1971; Enders and Schlafke, 1971; Steer, 1970a). Large granules which resemble secretory granules are frequently found with either electron translucent flocculous contents or with electron dense, fairly homogeneous contents which are partially connected to the endoplasmic reticulum. The cytoplasm contains abundant polysomes and free ribosomes. The uterine surface of the trophoblast is covered with single, upright, fairly long microvilli, which reach down deeply into the remnants of the blastocyst coverings which are undergoing dissolution. They appear embedded each in a pocket-like depression of the blastocyst coverings (Fig. 7). It is impossible to decide whether or not the light region surrounding the microvillus is an artifact of skrinkage or not. On the outer surface of the trophoblast, often at the base of the microvilli, we find micropinocytotic vesicles. A basal lamina underlies the trophoblast facing the blastocyst cavity.

The *entoderm* extends from the embryonic disc until not quite to the abembryonic pole so that it cannot be found everywhere in the obplacental region (Fig. 2). The entoderm cells which are stretched out extremely thin in many places contain, as do

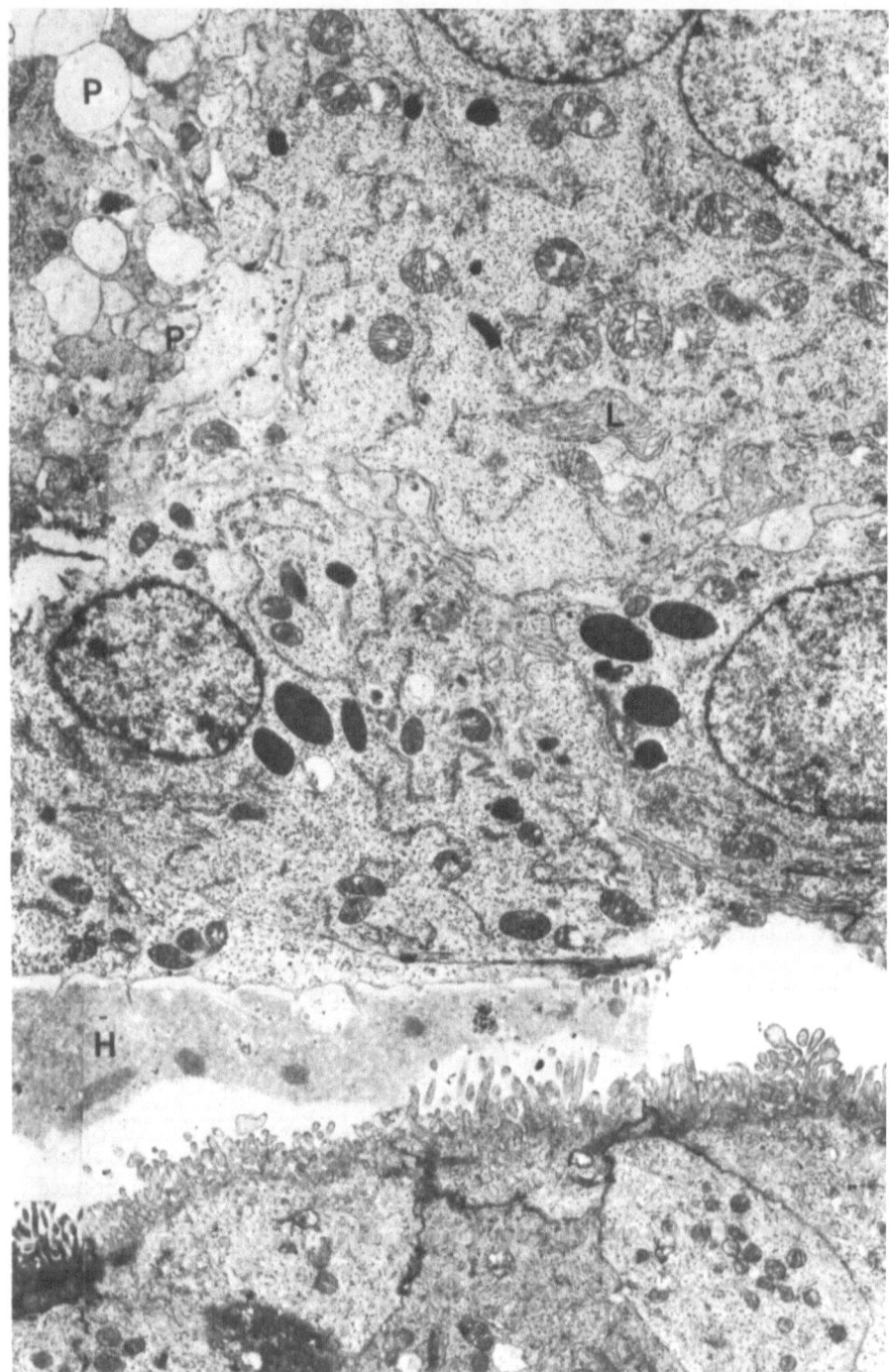

Fig. 7. Rabbit, normal implantation (control uterus, see 3.3.2.), morphology, 7 1/2 d p.c., anti-mesometrial-abembryonic region, EM X 6000. The trophoblastic knob seen here (above) has not yet attached to the uterine epithelium, but lies quite closely already. The blastocyst coverings (*H*) are undergoing dissolution. The microvilli of the surface of the trophoblast reach deeply into the swollen remains of the coverings, whereas the microvillous lawn of the uterine epithelial surface

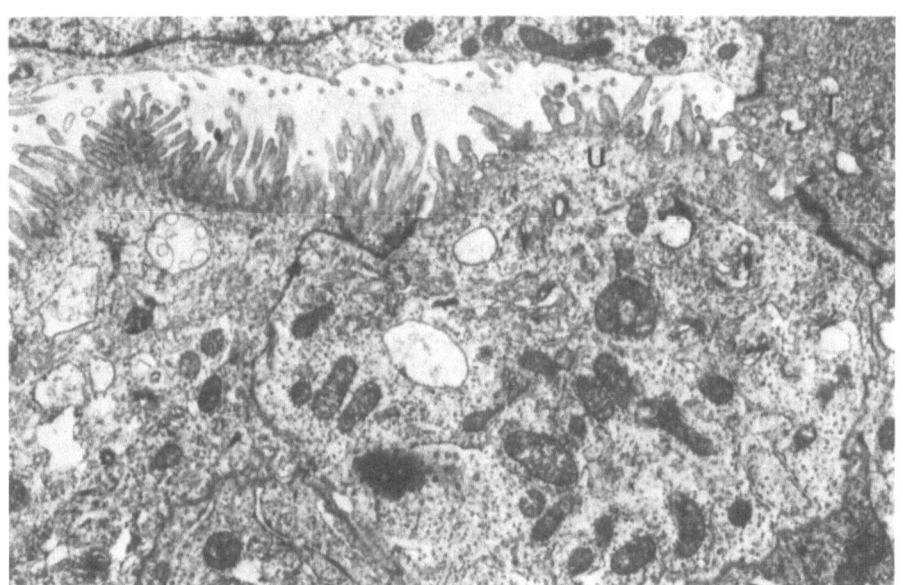

Fig. 8. Rabbit, normal implantation (control uterus, see 3.3.2), morphology, 7 1/2 d p.c., EM X 7000. The attachment of the trophoblastic knob *T* to the uterine epithelium *U* has begun (right in picture). In this region the adjoined cell surfaces are flat and the microvilli have disappeared here. In the left half of the picture the trophoblast and uterine epithelium are still some distance from each other; in places here the well-developed microvilli are loosely entwined. The uterine epithelial surface is (left in picture) enlarged by a peaked protuberance. The uterine epithelium still displays cell boundaries

the trophoblast cells, granules and vacuoles of varyingly dense content although those with electron translucent homogeneous contents which morphologically resemble lipid vacuoles predominate. In semi-thin sections the entoderm cells are characterized by their abundance of granules which can be strongly stained with toluidine blue. On the surface facing the blastocyst cavity the entoderm cells possess well-developed microvilli. Their endoplasmatic reticulum is often dilated to form wide cisterns (Fig. 24).

The *cavum epithelium of the uterus* shows a noticeably different behavior at the site of implantation than at the areas between the implantation sites. In the areas far away from the blastocysts (and in pseudopregnancy, see Larsen, 1962) the cells are multinucleate although most of the cell membranes remain intact. In the *region around the blastocysts*, on the other hand, nearly all cells of the cavum epithelium fuse and form wide symplasmatic discs, first antimesometrially and later (from 8 d p.c.) on top of the placental folds. At the time when the trophoblastic knobs make contact with the antimesometrial uterine cavum epithelium, the latter is still partially separated by cell borders and only later in the following half-day is the transformation to

---

remains separated from the latter by a space. The trophoblastic knob is partially syncytial (right top), in other regions cell boundaries can still be distinguished (below). It contains crystalloids (dark, oval structures) and packets of a labyrinthine membrane system (*L*). At the border towards the entoderm protuberances (*P*) are seen which appear to seal themselves off. The uterine cavum epithelium (below) still displays cell boundaries

wide symplasms completed (see Figs. 4, 6–8, 19, 23). Those parts of the uterine epithelium which lie directly over a subepithelial capillary probably play a special role in the antimesometrial *implantation* (Böving, 1962, 1963; Böving and Larsen, 1973): the trophoblastic knobs fuse selectively with such cells (Fig. 4). In the attachment phase of implantation, on part of the surface of the trophoblastic knobs a certain degree of *intertwining of the microvilli* with the very thick microvillous lawn of the uterine epithelium takes place (Fig. 8). There is, however, no regular, zipper-like interlocking over long stretches of the surface. *Fusion* also occurs at first only at small isolated spots at the surface of trophoblastic knobs involving single uterine epithelial cells. A trophoblastic knob can fuse with two or more epithelial cells which lie apart from one another and initially remain separated by other cells. In the beginning phase remnants of the cell membranes still show the plane of fusion. Cytoplasm and nuclei of the trophoblastic knobs then spread out in the direction of the subepithelial vessels (Fig. 4b) (Larsen, 1963; Enders, 1971, Enders and Schlafke, 1969, 1971; Schlafke and Enders, 1975). While this process is going on, the last lateral cell borders disappear in the antimesometrial uterine cavum epithelium. The trophoblastic knobs, which were at first only fused here and there with one or two uterine epithelial cells, now fuse laterally as well with the uterine symplasm. Some of the very flat trophoblast cells which lie between the trophoblastic knobs also fuse with each other during the next few days, although numerous cell borders still remain intact (cf. Fig. 23). These parts of the trophoblast mostly do not unite with the uterine epithelium either, but remain as bridges which are particularly noticeable over the uterine crypts. The epithelium deep in the antimesometrial *endometrial crypts* retains its cellular integrity.

At about 9 d p.c. the *obplacenta begins* to degenerate. The symplasms, which derive partly from the trophoblast and partly from the uterine epithelium, break away from the deeper parts of the endometrium. They develop vacuoles in the cytoplasm and pycnotic nuclei, die in the uterine cavity and are dissolved. The surface of the antimesometrial endometrium is once again covered with cellular epithelium, which grows out of remaining islands in the depths of the crypts. This process is completed by about 11 d p.c.

Mesometrial Implantation (Placentation)

The implantation of the blastocysts at the embryonic pole begins one days later, i. e. 8 d p.c. The histological details of this process differ in several respects from those described for abembryonic (antimesometrial) implantation. At the embryonic pole it is the trophoblast next to the embryonic disc which brings about the attachment in a zone which surrounds the caudal end of the embryonic anlage in horseshoe fashion (see Fig. 9). It is not separated into trophoplastic knobs but rather multilayered and transformed symplasmatically at the surface. The placental folds of the endometrium to which this part of the embryo is opposed proliferate greatly at this stage and have a highly vascularized stroma whose blood vessels herald the beginnings of differentiation of adventitial decidual cells especially in the deeper layers. On the surface of the placental area, but not laterally in the paraplacental fold or the endometrial crypts, the epithelium fuses to form wide symplasmatic plates. The cellular contact between embryo and mother comes about when *both symplasms (trophoblast and uterine epithelium of the placental folds) fuse with each other over large areas* (Larsen, 1961, 1970). It is not completely certain but usually assumed that the maternal nuclei die out in this common symplasm. Afterwards the trophoblast grows down further into the depths of the endometrium, and a haemochorial placenta is formed through erosion of the maternal blood vessels (Masquelin and Swaen, 1880; Duval, 1889a and b; Minot, 1889; Maximow, 1900; Schoenfeld, 1903; Mossman, 1926).

24

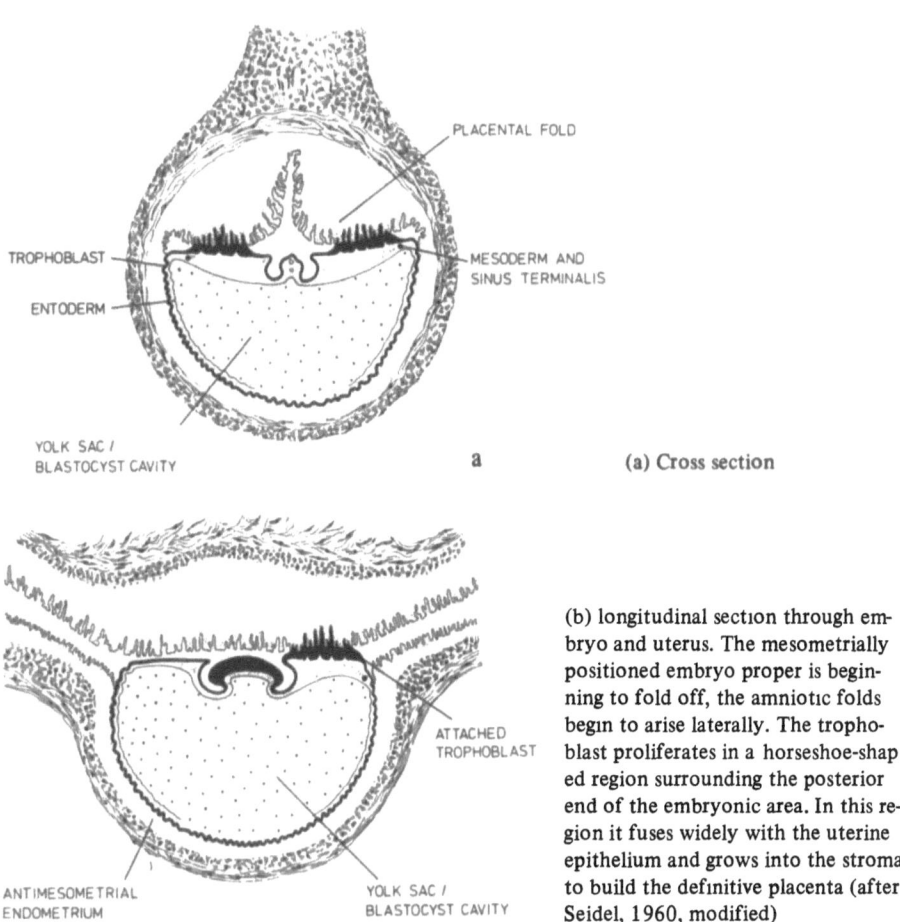

PLACENTAL FOLD

TROPHOBLAST

ENTODERM

MESODERM AND
SINUS TERMINALIS

YOLK SAC /
BLASTOCYST CAVITY

a

(a) Cross section

(b) longitudinal section through embryo and uterus. The mesometrially positioned embryo proper is beginning to fold off, the amniotic folds begin to arise laterally. The trophoblast proliferates in a horseshoe-shaped region surrounding the posterior end of the embryonic area. In this region it fuses widely with the uterine epithelium and grows into the stroma to build the definitive placenta (after Seidel, 1960, modified)

ATTACHED
TROPHOBLAST

ANTIMESOMETRIAL
ENDOMETRIUM

YOLK SAC /
BLASTOCYST CAVITY

Fig. 9a and b. Schematic drawing of the relative positions of embryo and uterus in the rabbit at the beginning of the embryonic (mesometrial) phase of implantation, 8 1/2 d p.c.

Endometrium

The light microscopical and electron microscopical morphology of the *endometrium* have already been exhaustively described for the rabbit in the early stages of pregnancy and pseudopregnancy (Beier, 1971, 1973, 1974a and b; Beier and Kühnel, 1973; Davies and Hoffman, 1973, 1975). However, only few descriptions exist in the work done by Larsen (1.c.) about the specific conditions in the *region around the blastocyst* at the site of implantation. The most remarkable feature of the endometrium in the implantation and early postimplantation stages is that the *cavum epithelium* in the blastocyst regions is changed into *wide symplasmatic plates,* on the placental folds as well as antimesometrially, to the extent that after 8 d p.c. lateral cell borders can rarely be found (compare Figs. 4, 6–8, 23, 24). Apparently the mechanism of this change is a cell fusion, as remnants of cell membranes with desmosomes can be found in the cytoplasm. The surface of this symplasm and of the still cellular antimesometrial crypts is enlarged not only by the development of a thick microvillous lawn but by pointed prominences covered with microvilli (Fig. 8). In the cavum epithelium lying between the implantation sites (as well as in the epithelium of the mesometrial and antimesometrial crypts) wide symplasms are not formed, but rather only a few cells fuse to form comparatively small multinuclear units.

3.1.2. Cat

The cat is, like the rabbit, a reflex ovulator, so that it is possible to determine the exact age of the embryos in this species. The development proceeds more slowly, however; all phases take at least twice as long as in the rabbit: ovulation 25–28 h p.c. (in the rabbit: around 10 1/2 h p.c.), transition from morula to the blastocyst stage and entrance into the uterus between 6 and 7 d p.c. (in the rabbit: 3 d p.c.), beginning of attachment in the uterus about 13 1/2 d p.c. (in the rabbit: 7 d p.c.), gestation an average of 65 days (in the rabbit: 30 days) (for literature see Denker et al., in preparation).

We have restricted our investigations to those stages where the attachment of the trophoblast to the uterine epithelium takes place. With the exception of work done at the end of the last century (done, however, more on the dog than on the cat; literature see Denker et al., in preparation) there are no detailed morphological accounts of these stages in the literature.

The investigations described here were done on unfixed cryostat sections as also used for enzyme histochemical protease assays (see 3.2.2.2.). While such sections offer few cytological details, they have the great advantage that no shrinkage artifacts and displacements need to be feared, such as those found with fixed and embedded material. Hence they are particularly appropriate for documentation of the topographical relationships between the still freely movable or attaching blastocysts and the uterine epithelium, as well as for judging the width of lumina (blastocyst cavity, cavum uteri including the slit between trophoblast and uterine epithelium, glandular openings). Likewise one need not fear an artifical lysis of the zona pellucida such as that which occurs with many fixatives (Denker, 1970a and b).

Morphology of the Uterus

A cat has the uterus bicornis typical for carnivores with a small corpus shared by both sides. In contrast to the rabbit, the endometrial glands are long, unbranching, tubulous and narrow. The epithelium of the deeper parts of the glands is distinctly different both morphologically and histochemically (see below) than that of the neck of the glands and than that of the cavum epithelium. The preimplantation phase brings about a development of the parts of the *endometrium near to the cavum,* which remotely recalls the transformation seen in the rabbit (Fig. 10). Whereas the endometrial surface is relatively smooth and interrupted only by the openings of the glands in the non-gravid animal, villi- and fold-like structures now begin to predominate (see Courrier and Gros, 1932a and b, 1933; Gros, 1933; Dawson and Kosters, 1944; dog: Barrau et al., 1975a). The transformation of the mucosa is by no means as highly developed as in the rabbit and it involves only those superficial parts of the endometrium nearest the cavum. These parts show a particularly high cell density in the stroma and a smaller size of the cells in proliferating epithelium (in connection with a larger nucleocytoplasmic ratio). The uterine *glands* grow in the preimplantation phase in length and their

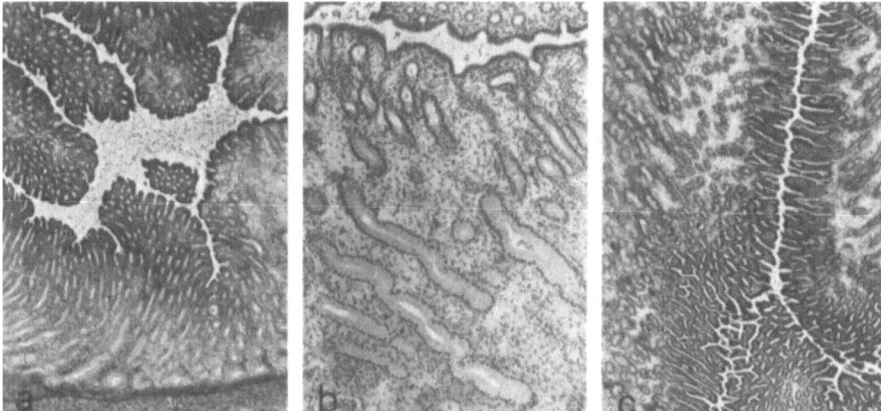

Fig. 10a–c. Cat, normal pregnancy, morphology of the endometrium, cryostat sections, HE.
(a) 13 p.c. X 20. The massive aggregation of leucocytes in the uterine cavity is typical for
uteri of the cat at this stage. (b) 13 d p.c. X 50. The lumina of the glands are relatively narrow
in the regions of the endometrium distant from the blastocysts. The glandular epithelium is co-
lumnar, the nuclei are positioned basally. Only in the necks of the glands is the epithelium isoprismatic. The stroma is rich in cells. (c) 14 d p.c. X 20. The superficial parts of the endometrium
have been transformed into irregular branches or villi-like structures. The lumina of the glands are
relatively narrow even at this stage in those regions of the uterus some distance from the blastocysts, and secretion plugs can be found in their necks

lumina widen. While during the early stages of pregnancy the lumina of the deeper
parts of the glands remain small and scarcely recognizable with the light microscope
and a plug of thick secretion builds up in the neck, *the glands begin to widen in the
direct neighborhood of the embryo at about 12–13 d p.c.,* while those in parts of the
uterus between the blastocysts change but little (Figs. 10, 15; Denker et al., in prep.).
This is an interesting morphological illustration of the fact that the blastocysts enter
into a physiological *exchange* with the surrounding endometrium before a cellular
contact has taken place (see also 4.1. and 4.6.). Characteristic for the sites of implantation is that the *secretion plug* of the neck of the endometrial glands *disappears*
mesometrially as well as antimesometrially (although not in the lateral aspect of the
implantation sites).

Morphology of Implantation

Implantation in the cat is, like that in the rabbit, of a central type. The feline embryo forms a
placenta zonaria, as is typical for carnivores. As with the paraplacental region, it has been mi-
croscopically well investigated (literature see Denker et al., in prep.). The feline placenta has a
lamellar construction and is endotheliochorial. There have been few detailed investigations done,
however, on the phases of the dissolution of the blastocyst coverings and on the first attachment of
the trophoblast to the endometrium, above all because of the difficulty mentioned in getting cats
to copulate at the time desired in the laboratory.

At the earliest stage we investigated, *12 d p.c.,* the slightly ellipsoid blastocyst (size
about 1.5 x 2 mm) still exists freely in the uterine lumen. At this time the extracellular
blastocyst coverings are being dissolved. In the cat these coverings are referred to as
the *zona pellucida* throughout their entire width, because there is no evidence here

(in contrast to the rabbit) for a deposition of tubal or uterine secretion. That the zona, in spite of their expansion, still surrounds the blastocysts until at least 11 d. p.c. can only be explained by the assumption that it has gained volume through swelling and/or the uptake of material (see 4.2.).

In the cat the blastocyst becomes oriented with the *embryonic disc* in the uterus *antimesometrially* (as opposed to the rabbit). This orientation is apparently not yet fixed at 12 d p.c. as the embryonic disc of the blastocysts investigated was found in varying positions, although mainly antimesometrially or antimesometrial-laterally.

In a part of the blastocysts investigated at 12 d p.c. the *zona pellucida* was dissolved on the abembryonic pole but still *existant over the embryonic disc.* In these cases the remaining ends of the zona were thinned-out, which points more to a *lysis* than a rupture. The thickness of the zona in certain blastocysts seemed reduced also over the embryonic disc, while the lateral regions are still intact.

The *trophoblast* cells in the cat are taller than in the rabbit, namely isoprismatic to fairly flat. The *embryonic disc* consists in this stage of a multi layered ectoderm and a well-developed entoderm which extends in the extraembryonic region as flattened cells.

The *uterine lumen* is packed full in this and the following stages with a great number of *polymorphonuclear granulocytes,* cell detritus and remnants of the glandular mucus plugs which can hardly be morphologically told apart from small remnants of the zona (Figs. 10, 13, 14). The amount of granulocytes is remarkably large, in complete contrast to the rabbit, where these elements can only be found rarely. In the direct neighborhood of the blastocyst all of these structures disappear nearly completely from the uterine lumen around 14 d p.c., which is possibly related to the lytic activity of the implanting embryo (see below).

13 d p.c., the blastocyst gains further in volume and usually takes on an elliptic form (size about 2 x 4 mm) (Fig. 14). It is not yet attached to the endometrium, although it lies in its smaller circumference closely against the uterine epithelium while between its lateral, pointed ends and the endometrium a space remains filled with uterine secretion. Blastocyst coverings can no longer be detected.

14 d p.c., the *invasion* has begun with all blastocysts. The blastocyst is greatly expanded (diameter about 5.5—6.0 mm) and forces the uterus outward so that the implantation sites are easily recognized from the outside. The blastocysts now lie closely apposed to the endometrium everywhere, laterally as well. In a belt-like region, which suggests the future placenta zonaria, the trophoblast appears under the light microscope to be fused with the uterine epithelium or the uterine epithelium is already largely replaced by the now layered trophoblast. In the laterally positioned, future paraplacentar region, a clear slit remains between the flatter single-layered trophoblast and the uterine epithelium, similar to that over the antimesometrially positioned embryonic disc. In these regions no evidence of an erosion can be seen on the uterine epithelium. In the region of the belt-like invasion zone, the uterine glands are greatly enlarged and free of secretion plugs (Fig. 15b, c). Laterally, in the future paraplacental region, they are, in contrast, still narrow-mouthed or only slightly enlarged and still contain secretion plugs at their necks.

3.2. Histochemistry and Biochemistry of Normal Implantation

3. 2. 1. Histochemistry of Mucosubstances at the Interface Between Embryo and Mother

Rabbit

Exhaustive substrate histochemical investigations have been carried out on the dissolution of the blastocyst coverings and the antimesometrial (obplacental) implantation in the rabbit (Denker, 1970a and b), and will only be recapitulated here in as far as they are essential to a discussion of the mechanism of implantation.

The *blastocyst coverings* in the rabbit are largely built of material which has been histochemically placed in the group of *acid mucosubstances* (MS) (polyanionic glycosaminoglycans and proteoglycans). The blastocyst coverings 6–7 d p.c. exhibit largely two *layers* in paraffin preparations; an intermediary, thin lamella can only be inconstantly demonstrated (see Denker, 1970a: Fig. 36a–d page 196). A clear separation into three layers can, however, be seen with the electron microscope and on semithin sections (see 3.1.1., Figs.19, 20). After histochemical investigations, an identification of these three layers with the coverings which already envelop the embryo upon entering the uterus (zona pellucida and mucoprotein layer) or with added uterine secretory material ("gloiolemma" of Böving, 1954, 1963) is more problematical than Böving originally supposed; this becomes especially clear when the existence of bound neuraminic acid (sialic acids) or sulfuric acid esters are considered: the zona pellucida contains sulfuric acid ester groups as well as terminal sialic acid. In the mucoprotein layer, on the other hand, we find sulfuric acid ester groups but, interestingly, no sialic acid (and none in masked form, for example esterised); the periodate-Schiff-reaction (PAS) reacts weakly as a result of the manifold esterification of the -OH groups with sulfuric acid. In the inner and outer layers of the blastocyst coverings, bound sialic acid can be demonstrated after 5–6 d p.c., and the PAS is strongly positive (Denker, 1970a). In these older blastocysts the outer layer is probably derived from uterine secretion ("gloiolemma"), while the mucoprotein layer forms probably the middle layer which can only bee seen clearly in semi-thin or thin sections, and the inner layer appears to be a new layer replacing the thinned-out zona pellucida (Denker and Gerdes, in preparation). The coverings of late preimplantation stage embryos thus are quite different from those of cleavage stages and young blastocysts.

The neuraminic acid which can be found in the coverings of older blastocysts is presumably largely 0-acetylated, as this MS becomes significantly neuraminidase-labile only after saponification (Denker, 1970a p. 197; Ravetto, 1968; Culling et al., 1974; Reid et al., 1976).

The sialic acid groups give the blastocyst coverings a certain *resistance to attack by proteinases,* which is theoretically plausible (see 4.3.). It was possible to show this on unfixed cryostat sections of 7 day rabbit blastocysts: after previous removal of the sialic acid by neuraminidase the proteolysis by trypsin was clearly facilitated (Denker, 1970a).

During the dissolution of the blastocyst coverings at the abembryonic pole of the blastocysts the individual chemical components seem to be differentially attacked by the enzymes involved. Thus the antimesometrially positioned, partially lysed remnants of the coverings are poor in neuraminic acid in comparison to the sulfate ester groups.

Apart from the blastocyst coverings, acid mucosubstances can be found in great quantities in a *surface coat of the uterine epithelium* representing a specifically well-developed glycocalyx (Denker, 1970a and b), and we find alternately neuraminic acid-rich and sulfate ester group-rich areas. A difference between the surface coat of the mesometrial and that of the antimesometrial endometrium can also be determined: in the region of the placental areas the neuraminic acid predominates (Denker, 1970a). As acid mucosubstances play an essential role in cell attachment processes, one may suspect that this difference has a meaning when taken together with the changes in the properties of the blastocyst coverings mucosubstances at the abembryonic pole during the *orientation of the blastocyst in the uterus* (with the abembryonic pole facing the antimesometrial endometrium).

Cat

The feline material on hand was used primarily for enzyme histochemical tests and therefore frozen unfixed. The results of investigations carried out on postfixed cryo-stat sections cannot be strictly compared with the results of paraffin sections. We have therefore restricted ourselves with the material taken from the cat to a few basically orientating substrate assays.

The PAS reacts strongly in the zona pellucida throughout its whole thickness at 12 d p.c. The secretion plugs in the necks of the uterine glands react equally strongly (see 3.1.2.) as well as their relatively sharply outlined fragments in the uterine lumen which probably are derived from them. 14 d p.c. such material cannot be found near the blastocyst either in the uterine lumen nor in the now greatly enlarged glandular lumina. The rest of the substances of the uterine secretion react fairly strongly positively in as far as they remain intact in the section after staining.

Reactions demonstrating acid MS (alcian blue pH 2.5 or pH 1.0, colloidal iron reaction after Hale) were very weak in the cryostat sections of uteri and blastocysts in the cat; only the blastocyst coverings (zona pellucida) and the surface coat of the uterine epithelium were fairly clearly stained.

3.2.2. Histochemical and Biochemical Enzyme Studies on Implantation Stages

3.2.2.1. Glycosidases

A number of glycosidases can be detected histochemically in the endometrium and trophoblast of the rabbit, which partially also appear in the uterine secretion (Fig. 35). As a result of known facts about the composition of the blastocyst coverings and the mucosubstances in the surface coat of the uterine epithelium (see 3.2.1.) we can expect that glycosidases play a role in the lysis of the coverings and in invasion. Some of these enzymes undergo characteristic changes in activity during the preimplantation phase, e. g. β-galactosidase (EC 3.2.1.23), β-N-acetylglucos-aminidase (EC 3.2.1.30), β-glucuronidase (EC 3.2.1.31) and α-amylase (EC 3.2.1.1) (Denker, 1971b).

3.2.2.2. Proteases

3.2.2.2.1. Aminopeptidases (Amino Acid Arylamidases; EC 3.4.11.2)

Rabbit

At least 3 enzymes of this category can be found in the uterus of pregnant rabbits and in the trophoblast. Although at least two of these enzymes have low substrate specificity, preliminary identification and differential localization of individual arylamidases can be chieved as follows (see Tab. 2, see also Denker and Stangl, 1976):

Table 2. Histochemical distribution of various arylamidases in trophoblast and uterine tissues in the rabbit during the preimplantation phase

	Trophoblast	Uterine epithelium	Blood vessels	Myometrium
Arylamidase I	+	+++ - ++++	+ - ++	+
Arylamidase II	++ - +++	(+)	+	+++
Arylamidase III	+ - ++	++	+++	++ - +++

Arylamidase I: localized especially in uterine epithelium and uterine secretion, also in abembryonic entoderm after 9 d p.c.; relatively wide substrate spectrum, strong activity with LeuNA; somewhat selectively detectable with the naphthylamides (NA) of aromatic amino acids (phe, tyr, trp).

Arylamidase II: (presumably related to aminopeptidase A, see Glenner and Folk, 1961; McMillan et al., 1962): localization in myometrium and trophoblast; typical substrates: naphthylamides of aminodicarbonic acids (α-glu and α-asp).

Arylamidase III: strong activity in the walls of blood vessels, but not clearly distinguishable from activity in uterine epithelium, myometrium and trophoblast; relatively wide substrate spectrum, which includes naphthylamides of gly, ala, arg and lys. The reaction with ArgNA is strongly reinforced by the addition of Cl^-. Mäkinen and Paunio (1972) consider this characteristic for aminopeptidase B. However no further indications for the presence of this enzyme, which specifically splits ArgNA and LysNA, could be read from the distribution pattern for the various substrates.

Arylamidase I, the enzyme of the endometrium and of the uterine secretion which reacts especially well with LeuNA, is most interesting as it shows a definite dependence on the maternal *progesterone* level and can first be clearly demonstrated at the time of the passage of the blastocyst from the tube into the uterus. The enzyme then appears in considerable quantities in the *uterine secretion* (Denker, 1969; Petry et al., 1970; Beier et al., 1972a; Denker and van Hoorn, 1974; Denker, 1976c; van Hoorn and Denker, 1975). Quantitative biochemical tests show that the activity in the uterine secretion reaches a sharp maximum at 5 d p.c. (in the range of 1000 mU/mg protein), and then sinks again (Fig. 11).

Although at first assumed to be *leucine aminopeptidase* (LAP) (Denker, 1969, 1971c; Petry et al., 1970; Beier et al., 1971, 1972a) the enzyme is not identical with classic LAP but is better designated as an *amino acid arylamidase* (see 4.4.1.; van Hoorn and Denker, 1975; van Hoorn, unpublished data; Denker and Stangl, 1976). Arylamidase I differs from the classic LAP (cytosol aminopeptidase, EC 3.4.11.1) in pH optimum (about 8.0), substrate specificity (preference for amino acid arylamides, but hardly any activity with leucine hydrazide) and in its being activated by Co^{++}

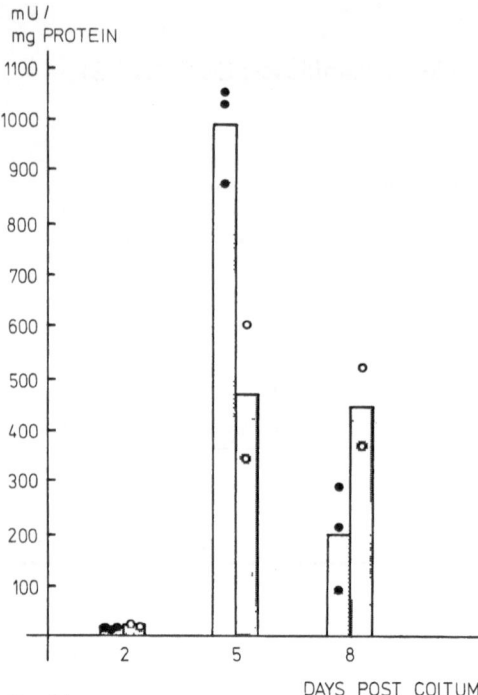

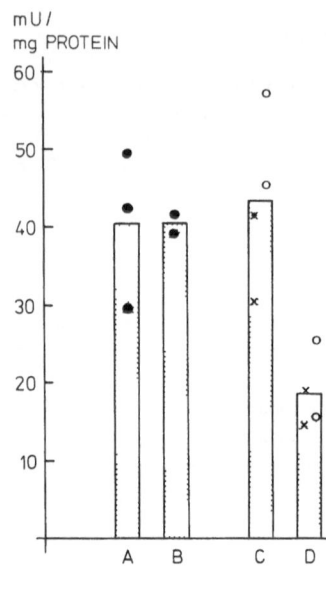

Fig. 11   Fig. 12

Fig. 11. Arylamidase activity in the uterine secretion of the rabbit in normal pregnancy and in pseudopregnancy. The dots or open circles mark values of one animal each, the height of the columns gives the mean value. Solid dots and lightly stipled columns: normal pregnancy. Values (mU/mg protein): 2 d p.c.: 18.6; 18.6; 10.7 ($\overline{x}$: 16.0). 5 d p.c.: 1051.9; 1029.5; 877.9 ($\overline{x}$: 986.4). Open circles and heavily stipled colums: pseudopregnancy. Values (mU/mg protein): 2 d p.c.: 20.2; 15.4 ($\overline{x}$: 17.8). 5 d p.c.: 600.8; 347.5 ($\overline{x}$: 474.1). 8 d p.c.: 366.9; 521.4 ($\overline{x}$: 444.1)

Fig. 12. Demonstration of an influence by the blastocyst on the endometrial arylamidase activity of the rabbit, 8 d p.c. The endometrium shows a lower activity at blastocyst sites D than in the region between the blastocyst sites (A and C) or than in pseudopregnancy B. The solid dots represent values for one animals each, the height of the colums gives the mean value. In C and D the symbols represent values of one uterus, and identical symbols are used for values fom the same animal.

A: Normal pregnancy; endometrium between the blastocyst sites (values from other animals than C/D) 49.2; 29.5; 42.2 ($\overline{x}$: 40.3) (mU/mg protein). B: Pseudopregnancy; values: 39.0; 41.6; ($\overline{x}$: 40.3) (mU/mg protein). C: Normal pregnancy; endometrium between the blastocyst sites (same animals as in D); values (mU/mg protein): animal I (x): 41.2; 30.2; animal 2 (o): 45.0; 57.0 ($\overline{x}$: 43.3). D: Normal pregnancy; blastocyst site endometrium, same animals as in C; values (mU/mg protein): animal I (x): 14.4; 18.6; animal 2 (o): 25,3, 15,3 ($\overline{x}$: 18,4)

as shown in biochemical tests. In histochemical test systems, the enzyme was found to be inhibited by EDTA and cysteine (10^{-2} M each), but not by iodoacetamide (10^{-2} M), Trasylol® (10^{-3} M) or SBTI (10^{-3} M).

Using improved histochemical methods (section freeze substitution) arylamidase I can be pinpointed to the *apical cytoplasm of the uterine epithelial cells* between 4 and 6 d p.c.; in the following stages the activity sinks in the parts of the endometrium lying between the blastocysts and in the pseudopregnant uterus, although only slowly, and the apical maximum disappears from the mesometrial uterine epithelium leaving a more uniform reaction of the whole cytoplasm. The sinking of the intracellular activity is greatly accelerated in the *neighborhood of the blastocysts*. This can be observed, interestingly, already at 6 2/3 d p.c., which is before the dissolution of the blastocyst coverings and before the beginning of attachment (van Hoorn and Denker, 1975). The histochemical observations are reinforced by quantitative biochemical tests, when the endometrium in the neighborhood of the blastocysts and the endometrium lying further away are separately tested (see Fig. 12). Because of the relative expense of experiments using pregnant rabbits only a few animals were tested biochemically. For this reason we must do without a statistical analysis. It becomes clear, nevertheless, that the parts of the endometrium near the blastocysts show a lower arylamidase activity than the other regions, especially when the corresponding values are compared for the same animal. This is impressively confirmed by the histochemical investigation of a larger series of animals, using longitudinal sections through the uterus (Denker, 1976c). The regions remote from the blastocysts apparently behave like the endometrium of pseudopregnant animals. The difference must be the presence or absence of blastocysts. The effect of the blastocyst cannot be ascribed to a release of arylamidase inhibitors: if homogenates of blastocysts and of endometria are combined, the activities behave additively, thus not giving any evidence for the presence of inhibitors. Apparently, through some mechanism as yet unknown, the *blastocysts stimulate the release of arylamidase from the uterine epithelium:* a comparison of the arylamidase activity in the uterine secretion of normally pregnant and pseudopregnant females shows that in the presence of blastocysts (normal pregnancy) the peak of activity is higher and is reached sooner (Fig. 11). This can best be understood as a result of a stimulation of the release of enzyme out of the endometrium. An inhibition of the synthesis de novo of arylamidase therefore seems improbable.

The nature of the signal which comes from the blastocyst is unknown. A purely mechanical effect of the embryo on the surrounding endometrium cannot be the mechanism: beads the same size as the blastocysts (see 2.1.) if placed in the uterus 6 d p.c., do not have a comparable effect on the endometrial arylamidase. On the other hand a local decline in the arylamidase activity has been observed near *copper IUDs* (Denker, 1976d; Denker and Kühnel, 1977). Biochemical tests show that copper ions strongly inhibit the uterine secretion arylamidase: complete inhibition by 1×10^{-3} M Cu^{++}; 30 % inhibition by 1×10^{-4} M Cu^{++}; at a Cu^{++} concentration of 1×10^{-5} M no further inhibition could be detected.

Cat

In feline uterine epithelium and secretion no arylamidase activity could be detected histochemically at 12–14 d p.c. using L-leucine-4-methoxy-β-naphthylamide. The trophoblast, however, gives at these stages a clearly positive reaction which may suggest that the enzyme has already been synthesized at this stage by the embryo and is not taken up from the maternal milieu. Presumably some enzyme released by the trophoblast is responsible for the slight reaction at the surface of the uterine epithelial cells near the trophoblast. The enzyme can be regarded as metal dependent, as it is completely inhibited by 10^{-2} M EDTA. There is, however, no evidence for any importance of SH-groups for the enzyme (no inhibition by 10^{-2} M iodoacetamide; no clear activation by 10^{-2} M cysteine, but no inhibition either).

3.2.2.2.2. Endopeptidases (Proteinases)

Histochemistry

Rabbit

The blastocysts of the rabbit develop a considerable proteolytic activity in the implantation stage, which can be well assayed and localized with a histochemical gelatin substrate film test (Denker, 1969, 1971d, 1972, 1974a and b, 1976a). A far lower activity of gelatinolytic enzymes is demonstrable by the same method in the uterine secretion. Because of its proven characteristic phase dependence and the localization of the enzyme activity, a special significance has, since the time it was first described, been suspected for the process of implantation. Young rabbit blastocysts show no activity in substrate film tests; 5–6 d p.c. such activity can be sporadically detected. But shortly before the beginning of the dissolution of the blastocyst coverings, 6 2/3 d p.c., a dramatic rise in the proteinase activity takes place at the surface of the blastocyst, whereby only the region around the embryonic disc remains behind.

We designate this enzyme or enzyme system as *"blastolemmase"*. As we hope to show in this paper, it apparently plays a central role in the dissolution of the blastocyst coverings (blastolemmata, see 3.1. and 4.2.). The activity can best be detected histochemically in the *blastocyst coverings,* while the *trophoblast* only hints at a reaction in the not experimentally tampered-with pregnancy (Kirchner, 1972a; Denker, 1974a, 1975). Biochemical tests, however, show a latent high activity in the trophoblast (see below). The maximal activity is reached between 7 and 7 1/2 d p.c., and is largely restricted to the abembryonic-antimesometrial region of the blastocyst, that is the region where the dissolution of the blastocyst coverings is under way (Figs. 18a, 32a). The main activity is found in the narrow space between trophoblast and uterine

Table 3. Histochemical determination of endopeptidase activity in the rabbit (Gelatin substrate film test, unbuffered, incubation period 1 3/4 h)

Localization	5d	6d	7d	7 1/2d	8 1/2d	9 1/2d	11 1/2d
Trophoblast							
abembryonic pole (blastolemmase)	φ	φ	φ - (+)	φ - +	φ - (+)	φ	φ
trophoblastic knobs (blastolemmase)			φ - +	φ - +	φ - (+)	φ	φ
embryonic disc region (blastolemmase)	φ	φ	φ	φ	φ	φ	φ
Blastocyst coverings							
abembryonic pole (blastolemmase)	φ - (+)	(+) - +	++	+++			
embryonic disc region (blastolemmase)	φ - (+)	(+)	(+) - +	φ - (+)	φ - (+)		
Entoderm							
(cathepsin)	φ	φ	(+)	+	++	++	++
Uterine secretion	(+) - +	(+) - +	+	+	+	+	(+) - +

epithelium and spreads to the glandular lumina and the regions of the uterine lumen near the blastocysts. The region of the embryonic disc which contains no trophoblast (Rauber's layer has already disappeared) gives a completely negative reaction as does the surrounding trophoblast of the embryonic third. The remnants of the blastocyst coverings which lie at this pole, which are at first not dissolved yet, only hint at an activity. They are pushed away from the blastocyst into the paraplacental furrow by the movements of the uterine tissues against the embryo. Here they will slowly be dissolved in the course of the next 1 1/2 days without showing a rise in proteinase activity. At 8 d p.c. when the antimesometrial implantation (obplacentation) is largely complete, only a slight remnant of blastolemmase activity can be found in the ab-embryonic-antimesometrial region, once again between trophoblast and uterine epithelium (glands). This finding is typical for all stages that we investigated until 11 1/2 d p.c.

At 7 1/2 d p.c. a proteinase activity begins to appear in the *entoderm* of the embryo, which can be attributed to a different cathepsin-like enzyme (see Tab. 3, 16, and 4.4.2.1.). It first occurs in the region of the embryonic disc, later especially in the entoderm directly next to the embryonic disc and spreads eventually widely laterally until 11 1/2 d p.c. (Fig. 22a, c, 26b). At 9 1/2 d p.c. the reaction is noticeably restricted to extraembryonic entoderm while the entoderm of the embryonic disc shows no activity. The extraembryonic entoderm shows greatest activity in those regions which lie beneath the mesoderm. This enzyme cannot be detected histochemically in the mesoderm itself nor in the ectoderm nor can it be found in the amnion or the allantois.

The *blastocyst fluid* remains free of significant activity at all stages.

At 9 1/2 d p.c., in the *endometrium*, particularly on the mesometrial side, a proteinase activity arises which can be clearly distinguished from blastolemmase by its resistance to the various inhibitors (see below). It can be demonstrated well in the deeper parts of the lumen of the endometrial crypts of the placental folds and, to a lesser extent, also on the surface of the cavum epithelium and distributed diffusely in subepithelial stroma (also between implantation sites). In the endometrial *stroma* and in the myometrium, proteinase positive cells can be found; their numbers increase clearly mesometrially as well as antimesometrially after 7 1/2 to 8 d p.c., especially at the site of implantation. This enzyme is also biochemically different from blastolemmase, although related to or identical with the cathepsin-like proteinase of the entoderm (see below, 4.4.2.1.).

Cat

In the earliest stages of pregnancy which we investigated (*12 d p.c.*) the gelatin substrate film test showed only weak to fair endopeptidase activity in most parts of the uterus and in the blastocysts (Tab. 4). At the surface of the *uterine epithelium* and in the *uterine secretion* (in the uterine lumen and the mouths of the glands) we find a proteinase reaction remarkably regularly in the regions remote from the embryo, but hardly at all in the neighborhood of the blastocysts. The deeper parts of the endometrial glands show no reaction; the middle regions, like the cavum epithelium, evince an activity which arises from largely basally located intracytoplasmatic granules or vesicles (lysosomes?). Similar focal reactions take place in the *trophoblast*, with, however, no special preference for certain regions of the cytoplasm. Occasionally the reaction predominates near the outer surface of the trophoblast cells and between them

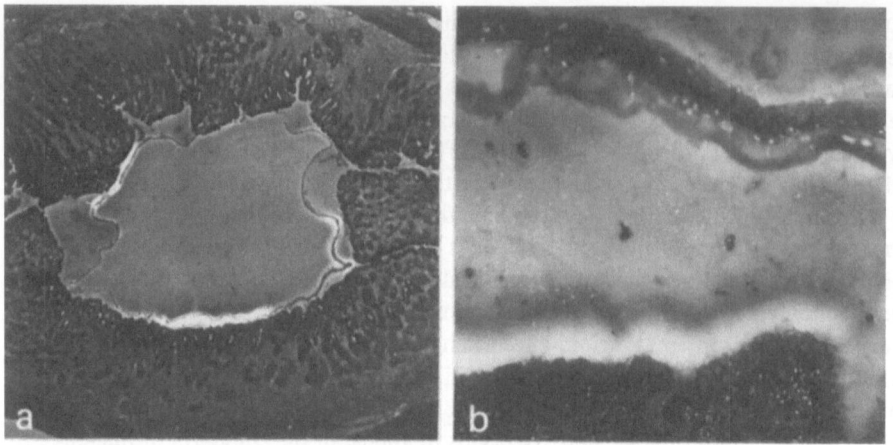

Fig. 13a and b. Cat, normal pregnancy, proteinase histochemisty, 12 d p.c., cryostat sections, gelatin substrate film test.

(a) Blastocyst in the uterus, incubation period 2 1/2 h, X 15. The walls of the blastocyst show fairly strong proteinase activity in spots. Endometrium and uterine secretion show no significant reaction. (b) Part of the trophoblast, zona pellucida and endometrium, incubation period 6 1/2 h, X 260. Shown here is a region of relatively little proteinase activity. At this high magnification we see that a focal reaction occurs in the trophoblast (above). The individual foci lie strewn irregularly in the cytoplasm, but often at the border between trophoblast and the zona pellucida undergoing dissolution. Focal reactions also occur in the uterine epithelium (below), but are less obvious there. A diffuse zone of lysis is recognizable on the surface of the cavum epithelium, however. Leucocytes in the uterine cavity show usually only little activity.

Table 4. Histochemical determination of endopeptidase activity in the cat (Gelatin substrate film test, unbuffered, incubation period 1 1/2 h)

Localization	12 d	13 d	14 d
trophoblast	+ (- ++)	++	++[a] - ++++[b]
embryonic disc region	ϕ - (+)	ϕ	ϕ
zona pellucida	(+)		
uterine secretion			
near embryo	ϕ - (+)	(+)	(+)[a]
remote from embryo	(+)	(+) - +	+
endometrial glands			
near embryo	ϕ - (+)	(ϕ) - +	++ - +++
remote from embryo	ϕ - +	+	+

[a] outside of the invasion zone
[b] invasion zone

and the zona pellucida (Fig. 13b). The *zona* and its dissolving remnants show little activity in the gelatin film test, as do the numerous leucocytes lying in the uterine lumen. In places where the trophoblast lies closely to the uterine epithelium we find a greater activity, especially at the surface of the uterine epithelium and in the space between it and the trophoblast.

36

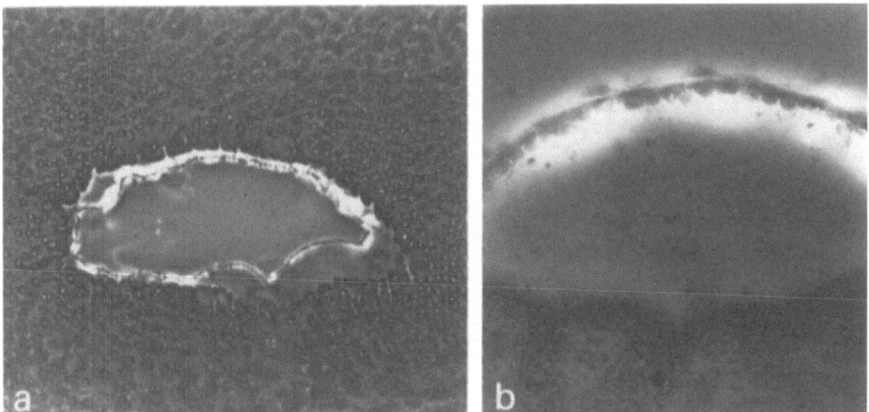

Fig. 14a and b. Cat, normal pregnancy, proteinase histochemistry, 13 d p.c., cryostat sections, gelatin substrate film test.
(a) Blastocyst in uterus, incubation period 4 h, X 25. Blastocyst sectioned somewhat peripherally. Clear proteinase activity originates in the trophoblast. The surrounding endometrium shows no significant reaction after this incubation time. (b) Section of the blastocyst wall and endometrium, incubation period 3 1/4 h, X 170. The trophoblast cells are isoprismatic. They show high proteinase activity. Entoderm (flat epithelium) is seen above. Numerous polymorphonuclear granulocytes in the uterine cavum

13 d p.c. the zona pellucida is dissolved, the blastocyst is further expanded and lies closer to the uterine epithelium but is not yet attached (see 3.1.). The substrate film test gives a largely similar picture to that described for 12 d p.c., but the activity in the trophoblast is greater (Fig. 14).

14 d p.c., at the beginning of invasion, a considerable proteinase activity is found at the site of implantation, which is significantly greater than that of the rabbit blastocyst. Lysis zones of a width comparable to those reached by the 7 1/2 days old rabbit blastocyst after 1 1/2—2 hours are reached here in 15 min. This high activity is restricted to the belt-shaped *region of invasion* (Fig. 15a) and can be found in the neighborhood of the trophoblast, the uterine epithelium under destruction and the widened glands. The activity is so great there that a certain amount of auto-digestion of cells takes place in the sections even at room temperature. The flatter trophoblast of the lateral parts of the blastocyst which lies only loosely against the uterine epithelium shows no activity in some areas and in others shows a proteinase activity reminiscent of the 13 day stage. The enzyme activity is greater in the *endometrium and uterine secretion between the implantation sites* than in the neighborhood of the embryo (apart from the above mentioned high activity in the invasion zone). The cytological details of localization have not changed greatly compared to the stage 12 d p.c., although the activity is overall increased.

No proteinase activity was observed in the *blastocyst fluid* at any stage we investigated.

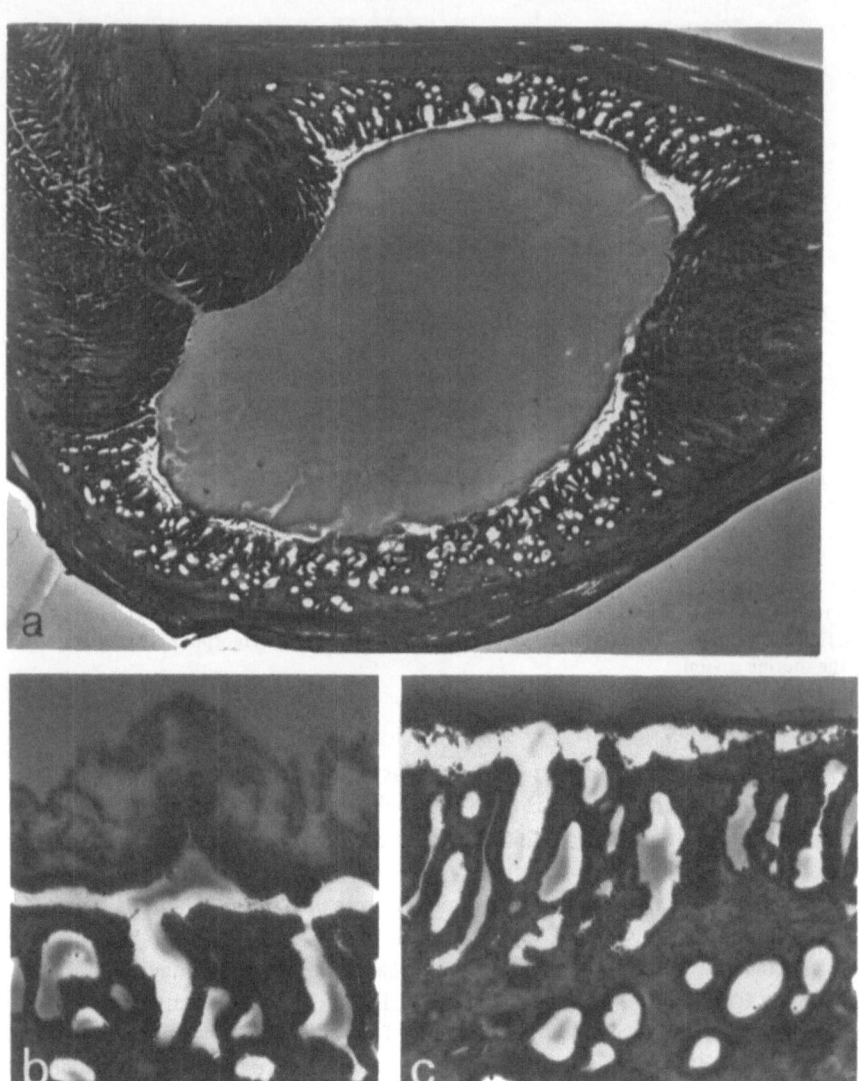

Fig. 15a–c. Cat, normal pregnancy, proteinase histochemistry, 14 d p.c., cryostat sections, gelatin substrate film test.

(a) Blastocyst in uterus, incubation period 25 min X 12. The uterus has been sectioned longitudinally. In the girdle-like zone of invasion (see above and below) very high proteinase activity can be detected (very short incubation time!). The presumptive paraplacental trophoblast (right and left) has much less activity which is only detectable after longer incubation periods as is also true for interblastocyst parts of the endometrium. (b) Embryonic disc and antimesometrial endometrium, incubation period 10 min, X 70. The neural groove, well developed mesoderm and entoderm can be recognized in the embryonic area (from bottom to top). No significant proteinase activity can be detected in these tissues. It is confined to the area between the embryonic area and the uterine cavum epithelium and to the antimesometrial endometrial glands (below) with their greatly widened lumina. (c) As in Fig. 15b, invasion zone, incubation period 15 min. The trophoblast of the invasion zone erodes the uterine cavum epithelium. The maximum of proteinase activity is found in this region. Following lysis the gelatin substrate film has retracted from this spot together with the tissues leaving an artificial space between endometrium and trophoblast. Some proteinase activity can also be found in the greatly widened lumina of the endometrial glands

38

Attempts at a Biochemical Characterization of the Blastocyst and Uterine Secretion Proteinases

Characterization of the Active Site of Blastocyst and Uterine Secretion Proteinase by in vitro Inhibition Experiments

Rabbit

In a large series of experiments the interaction of inhibitors with proteinases of the blastocyst and of the uterine secretion in the rabbit in vitro was studied using the substrate film test. In this way we persued two goals: 1. preliminary classification by characterization of the active site of the enzymes; 2. preparation of in vivo inhibition experiments (see 3.3.2.).

Active site-directed inhibitors were preferably tested. As source of the enzymes we used cryostat sections from 7 d p.c. rabbit blastocysts in the uterus.

Table 5 shows those inhibitors and their concentrations (concentrations as present in the solutions used for the soaking of the gelatin substrate film) which will inhibit the proteinase activity of the *trophoblast* and the *blastocyst coverings (blastolemmase)*. The inhibitor effects are qualitatively the same at various pH values. In the case of DFP it must be noted that this inhibitor evaporates rapidly so that it is not possible to keep a defined amount on the substrate film. The concentration values given are therefore possibly too high. The chloromethylketones (TLCK, TPCK) are poorly adapted to these tests where a preincubation of the enzyme with the inhibitor is technically im-

Table 5. Inhibition of the trophoblast-dependent blastocyst proteinase (blastolemmase) in the rabbit by various proteinase inhibitors in vitro
(Histochemical gelatin substrate film test, stage tested: 7 d p.c.)

Inhibitor	Specificity[a]	Lowest still clearly inhibiting concentration (M)
1. SBTI	Try, Chy	$1 \cdot 10^{-6}$
2. Trasylol®	Try, Chy	$1 \cdot 10^{-6}$
3. antipain	Try	$5 \cdot 10^{-6}$
4. leupeptin	Try	$5 \cdot 10^{-6}$
5. SSPI I + II	Try	$5 \cdot 10^{-6}$
6. PSTI	Try[b]	$1 \cdot 10^{-5}$
7. α_1-antitrypsin	Try, Chy	$1 \cdot 10^{-5}$
8. NPGB	Try	$5 \cdot 10^{-5}$
9. α_1-antichymotrypsin	Chy	$1 \cdot 10^{-4}$
10. ovomucoid (chicken)	Try	$5 \cdot 10^{-4}$
11. chymostatin	Chy	$1 \cdot 10^{-3}$
12. p-aminobenzamidine	Try	$1 \cdot 10^{-3}$
13. DFP	Try, Chy	$1 \cdot 10^{-3}$ (?)
14. TLCK	Try	$1 \cdot 10^{-2}$

[a] only listed in regard to trypsin or chymotrypsin inhibition
[b] except human trypsin

Literature regarding specificity and properties of inhibitors see Denker (1976a). Additional references on inhibitors no, 2, 3, 5 and 8 see chapter 4.4.2.2.1

Table 6. Inhibitors which do *not* inhibit the trophoblast-dependent blastocyst proteinase (blasto-lemmase) in the rabbit in vitro

(Histochemical gelatin substrate film test, stage tested: 7 d p.c.)

Inhibitor	Specificity[a]	Highest concentration tested (M)
1. EACA	[d] (Try)	$1 \cdot 10^{-2}$
2. AMCHA	[d] (Try)	$5 \cdot 10^{-1}$
3. pepstatin	[b]	$5 \cdot 10^{-2}$
4. DSI	Try, Chy	$1 \cdot 10^{-3}$
5. GPNA[c]	Chy	$1 \cdot 10^{-2}$
6. BANA[c]	Try	$1 \cdot 10^{-2}$
7. TPCK	Chy	$1 \cdot 10^{-2}$
8. NBD	Chy	$1 \cdot 10^{-2}$
9. KCN		$1 \cdot 10^{-2}$
10. EDTA		$1 \cdot 10^{-2}$
11. iodoacetamide		$1 \cdot 10^{-2}$

[a] given only in relation to trypsin and chymotrypsin inhibition
[b] pepsin, cathepsin D
[c] substrate, tested here for ability to inhibit competitively
[d] plasminogen activation (plasmin)

For literature regarding specificity and properties of inhibitors see Denker (1976a). For additional references on inhibitors no. 1 and 2 see Andersson et al. (1965); Bang (1971); Umezawa (1972); Witt (1975). No. 4 see Fritz et al. (1971); Fritz and Hochstrasser (1976). No 8 see Kézdy and Kaiser (1970)

possible (see Denker, 1971c), as with such substances the complex formation takes place too slowly.

Table 6 shows which inhibitors gave no inhibition of the blastolemmase.
Cysteine (10^{-2} M) does not markedly influence the blastolemmase activity either in the sense of an inhibition nor of an activation.

The *entoderm proteinase* differs from blastolemmase clearly in that it is only inhibited by antipain, chymostatin and iodoacetamide (Table 7). Inhibition by NBD can only be observed when phosphate buffer is used. As phosphate buffer catalyzes the spontaneous hydrolysis of NBD (see Kézdy and Kaiser, 1970), caution must be used in the evaluation of results. The typical trypsin inhibitors show no inhibitor effect. Cysteine (10^{-2} M) activates this enzyme.

The proteinase activity of the *stroma cells* exhibits the same spectrum of effective inhibitors as the entoderm proteinase (see Tab. 7). The proteinase which is at 9 1/2 d p.c. more strongly detectable in the depths of the *endometrial crypts* and at the surface of the cavum epithelium behaves similarly. An activation by cysteine (10^{-2} M) can also be clearly detected here.

The reaction of the *uterine secretion* varies strongly, so that an evaluation of the effect of inhibitors is made difficult. Trasylol®, leupeptin, antipain, PSTI, NPGB, ovomucoid, SBTI and SSPI II operate as inhibitors. Weak, incomplete inhibition is seen with NBD. No inhibition was detected in the case of pepstatin, chymostatin and EDTA.

Table 7. Inhibition of the entoderm proteinase and the stroma cell proteinase in the rabbit by various proteinase inhibitors in vitro

(Histochemical gelatin substrate film test, stages tested: 7–11 d p.c.)

Inhibitor		Entoderm proteinase	Stroma cell proteinase
1. iodoacetamide	10^{-2} M	+	+
2. antipain	10^{-3} M	+	+
3. chymostatin	10^{-3} M	+	+
4. NBD	10^{-2} M	(+)	(+)
5. NPGB	10^{-2} M	ϕ	ϕ
6. Trasylol®	10^{-3} M	ϕ	ϕ
7. ovomucoid (chicken)	10^{-3} M	ϕ	ϕ
8. SBTI	10^{-4} M	ϕ	ϕ
9. pepstatin	10^{-3} M	ϕ	ϕ
10. SSPI – II	10^{-4} M	ϕ	
11. DSI	10^{-4} M	ϕ	ϕ
12. EDTA	10^{-2} M	ϕ	ϕ

For literature regarding specificity of inhibitors see legends to Tab. 5 and Tab. 6

Cat

The proteinases of the *trophoblast* and the *uterine secretion* in the cat (as opposed to in the rabbit) are resistant to most of the inhibitors tested (see Tab. 8). Iodoacetamide and chymostatin inhibit. The weak remnants of activity which the trophoblast exhibits at pH 8 can be inhibited by antipain, although this inhibitor shows an uncertain effect on the strong reaction at pH 5.0 or on the uterine secretion. NBD gives a hint of inhi-

Table 8. Inhibition of trophoblast and uterine proteinases in the cat by various proteinase inhibitors in vitro

(Histochemical gelatin substrate film test, stages tested: 12 and 14 d p.c.)

Inhibitor		Trophoblast	Uterine epithelium, uterine secretion (pH 5)	Stroma cells (pH 8)
1. iodoacetamide	10^{-2} M	+	+	ϕ
2. antipain	10^{-3} M	(+)	ϕ - (+)	+
3. chymostatin	10^{-3} M	+	+[a]	ϕ
4. NBD	10^{-2} M	(+)	ϕ	(+)
5. NPGB	10^{-2} M	ϕ - (+)	ϕ	+
6. Trasylol®	10^{-3} M	ϕ	ϕ	+
7. ovomucoid (chicken)	10^{-3} M	ϕ	ϕ	+
8. SBTI	10^{-3} M	ϕ	ϕ - (+)	+
9. pepstatin	10^{-3} M	ϕ	ϕ	ϕ
10. EACA	10^{-2} M	ϕ	ϕ	ϕ
11. EDTA	10^{-2} M	ϕ	ϕ	ϕ

[a] locally incomplete

Regarding specificity of inhibitors see legends to Tab. 5 and Tab. 6

bition of the activity of the trophoblast. Cysteine (10^{-2} M) here operates clearly as an activator, although this cannot be demonstrated in the uterine secretion.

The *stroma cell proteinase,* as opposed to in the rabbit, is inhibited by the typical trypsin inhibitors (see Tab. 7, 8).

Electrophoresis

The electrophoretic migratory behavior of the blastocyst and uterine secretory proteinase was investigated in the *rabbit* by *micro disc electrophoresis.* Proteinase zymograms obtained with the *gelatin* substrate film test are pictured schematically in Fig. 16. The statements made about the relation to the main protein bands rest on an evaluation of gels longitudinally sectioned on the cryostat, on which a proteinase reaction

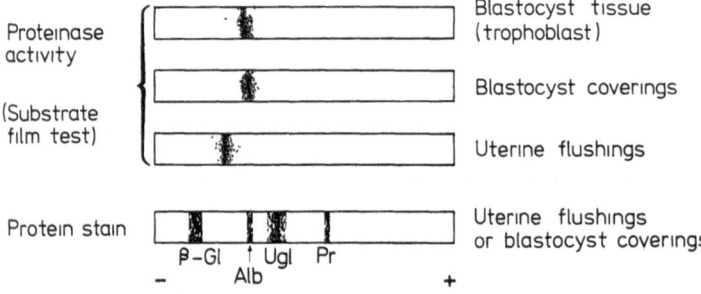

Fig. 16. Migratory properties of the gelatinolytic proteinases from blastocyst tissue (trophoblast), blastocyst coverings and uterine flushings in the rabbit in microdisc electrophoresis. Approximate Rf-values: proteinases: blastocyst tissue (trophoblast) 0,30; blastocyst coverings 0.33; uterine flushings 0.24. Major proteins: β-glycoprotein (β-Gl) 0.14; albumin (Alb) 0.32; uteroglobin (Ugl) 0.42; prealbumin (Pr) 0.58

as well as a proteinase staining was carried out. The localization of the proteinase fractions was not very sharp. The Rf-values given are therefore only to be taken as an approximation (number of electrophoreses: uterine secretion: 19; blastocyst tissue: 15; blastocyst coverings: 11).

In trophoblast, blastocyst coverings and uterine secretion, it is always only one fraction which shows significant gelatin-splitting activity. It is in any case clearly separated grom β-glycoprotein. The proteinase from the *trophoblast* and the *blastocyst coverings* migrate close to albumin and are difficult to separate from it. There appears to be a certain difference in migratory behavior between *uterine secretion proteinase* on the one hand, and blastocyst tissue and coverings proteinase on the other hand although this difference cannot be determined with certainty using this method.

As an alternative to mounting the gels on substrate films an attempt was made at including gelatin in the PAA-gel. This procedure so far did not give satisfactory results for micro disc electrophoresis. When such electrophoresis in gels containing gelatin was performed at macroscale, the blastocyst and the uterine secretion proteinases migrated very poorly at alkaline pH, and were clearly separable also in this case from β-glycoprotein. A separation at acid pH, as was used for trypsin, was not possible as the proteinases under investigation were found to be unstable under these conditions.

Various fractions react after micro disc electrophoresis with the *synthetic substrates* BANA and GPNA. They are largely not identical with those which hydrolyze gelatin. At pH 8.0 all reactions are much stronger than at pH 6.0, but it must be noted that

GPNA dissolves poorly at pH 6.0. With BANA the *uterine secretion* shows a strong band in the zone active in the gelatin test as well (Rf 0.25–0.32. The high dye binding affinity of the neighboring albumin possibly shafts the band artificially towards this fraction). This fraction reacts weakly or not at all with GPNA. It can be found, although faint, in the uteri of pseudopregnant females. It can be well separated from arylamidase I (see 3.2.2.2.1.): the strong band obtained with LeuNA has an Rf of about 0.18. An additional, more slowly migrating fraction (Rf about 0.10) is stained lightly with both GPNA and BANA and is found in changing amounts in the normal pregnancy. *Trophoblast* and *blastocyst coverings* show several only slightly positive bands in the region between the start point and the region of the β-glycoprotein (Rf-value between 0.02 and 0.15). In the trophoblast the reaction with GPNA is stronger than with BANA. The fraction which reacts strongly in the gelatin test (*blastolemmase*) shows *no significant staining with BANA and GPNA*. In *entoderm* homogenates at pH 6.0 (acid optimum of the entoderm proteinase, see below) no bands which react notably with BANA or GPNA can be recognized.

The major fraction demonstrable with gelatin as a substrate migrates in *agar electrophoresis* to the anode in the case of blastocyst tissue and blastocyst coverings, and to the cathode in the case of uterine flushings of the same stage (6 2/3 d p.c.), although a fraction which wandered to the anode was found in some samples of uterine flushings (see Fig. 17). These flushings originated from uteri with blastocysts and could have been contaminated with blastocyst proteinases.

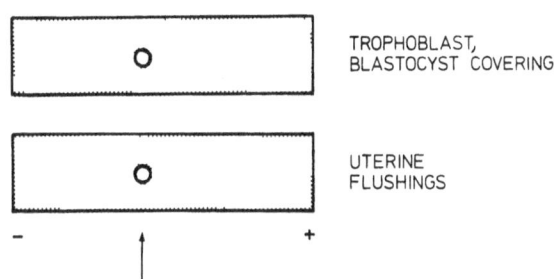

TROPHOBLAST,
BLASTOCYST COVERINGS

UTERINE
FLUSHINGS

− ↑ +

Fig. 17. Migratory properties of gelatinolytic proteinases from trophoblast (blastocyst tissue), blastocyst coverings and uterine flushings in the rabbit using agar gel electrophoresis. The arrow points to the start point. The fraction which migrates to the anode cannot be seen in all samples of uterine flushings

Substrate Specificity and Further Biochemical Properties of the Blastocyst and Uterine Secretion Proteinases

Substrate Specificity

The *trophoblast proteinase (blastolemmase)* of the *rabbit* shows no significant activity towards the usual synthetic trypsin and chymotrypsin substrates. This is all the more remarkable as the active site of the enzyme must have considerable similarities with trypsin and some relationship to chymotrypsin because of its spectrum of effective inhibitors (see the previous chapter and 4.4.2.1.). The following substrates were tested photometrically and found *poorly hydrolyzable* by blastolemmase:

Trypsin substrates: benzoyl-arginine ethyl ester (BAEE), benzoyl-arginine-β-naphthylamide (BANA), benzoyl-arginine-p-nitroanilide (BAPA).

Chymotrypsin substrates: acetyl tyrosine ethyl ester (ATEE), carboxypropionyl-phenylalanine-p-nitroanilide, glutarylphenylalanine-β-naphthylamide (GPNA).

Elastase substrates: elastin-orcein, succinyl-tri-alanine-p-nitroanilide. N-acteyl-alanine-p-nitroanilide.

The activity measured using casein as a substrate is very low. We used the azocasein test (see Fritz et al. 1974b), the dimethylcasein test (see LIN et al., 1969) and a test based on the cleavage of Hammarsten casein followed by Lowry test determination of TCA-soluble fragments released. Attempts to carry out a histochemical substrate film test using casein films (see Yamada, 1964) instead of gelatin films gave entirely negative results with sections of rabbit blastocysts. No hydrolysis of fibrin was observed when fibrin agar films (prepared according to Heimburger and Schwick, 1962, some of them excessively washed to remove remnants of hypophosphite) were tested in the same manner.

The best substrate found for this enzyme to date is *gelatin.*

A BANA and BAPA cleaving activity can be demonstrated in the *uterine secretion* of the rabbit. In micro disc electrophoresis it migrates in the same fraction as the gelatinolytic activity (see chapter electrophoresis). GPNA is cleaved only poorly by this fraction.

The *entoderm* proteinase of the rabbit appears not to cleave BANA and GPNA in demonstrable amounts (see electrophoreses).

In the *cat* the substrate specificity of the proteinases could only be tested histochemically. The proteinase activity of the trophoblast and the endometrial glands at the implantation site showed a clear preference for gelatin as a substrate; some hydrolysis of casein films, however, was indicated. The trophoblast (and, to a minor extent, the adjacent uterine epithelium also) shows high activity using L-leucine-4-methyoxy-β-napththylamide as a substrate but not when BANA or GPNA are used. This enzyme can be clearly distinguished from the gelatinolytic activity by its inhibition through EDTA (see 3.2.2.2.1.).

pH optimum

As no appropriate substrates for blastolemmase are known which are suitable for quantitative photometric tests, we had to resort to the histochemical substrate film test to investigate the pH dependency of the proteinase reaction. A determination of the pH optimum using this method can only be an approximation, as incubation in a liquid medium is not possible (see Denker, 1971c), but rather the substrate films had to be incubated with the mounted sections in a humid atmosphere. By soaking the substrate films before use in a non-volatile buffer (phosphate buffer) of the desired pH and drying them before mounting, a certain degree of pH control could be achieved. Using this procedure one obtains the following results of the pH dependency (see Tab. 9):

In the *rabbit* the reaction at the abembryonic pole of the blastocyst (blastolemmase) is markedly enhanced at alkaline pH (8.5), although it can already be seen at pH 5.0. Lateral regions and regions of the trophoblast nearer the embryonic pole give a more intense reaction at slightly acid pH values, which is also true for the activity of the trophoblastic knobs seen after in vivo treatment with antipain (see 3.3.2.1.). In the uterine secretion the pH optimum lies in the more alkaline region, whereas entoderm and stroma cell proteases act preferably at more acid pH. The proteinase activity, which one finds after 8 1/2 d p.c. in the depths of the endometrial crypts and at the uterine epithelial surface, also has its maximum at slightly acid pH values.

In the *cat* 12 d p.c. one finds a poorly defined optimum on the trophoblast surface at neutral pH (or slightly basic pH), but 14 d p.c. the optimum lies in the acid region without, however, being completely suppressed when the pH is raised to values around pH 8–8.5. At the surface of the uterine epithelium and in the glandular lumina the reaction is strongest at acid pH; in the case of the stroma cells, as opposed to the situation in the rabbit, it is strongest at basic pH (Tab. 9).

Table 9. pH-dependence of blastocyst and uterine proteinases in the rabbit and the cat (Histochemical gelatin substrate film test)

	pH 5.0	6.0	7.0	8.0	8.5
Rabbit					
trophoblastic surface/ blastocyst coverings, abembryonically (7 1/2 d p.c.)	++	++	++ - +++	+++	+++ - ++++
entoderm (9 1/2 d p.c.)	++	+ - ++	+	(+)	φ
uterine secretion (7 1/2–9 1/2 d p.c.)	φ - (+)	φ - (+)	(+) - +	+	+ (- ++)
stroma cells (7 1/2–9 1/2 d p.c.)	+ - ++	+	(+)	φ	φ
Cat					
trophoblast					
12 d p.c.	φ	(+)	+	(+) - +	(+)
14 d p.c.	+ - ++	(+) - ++	++	+	φ - +
uterus: cavum epithelium (12 d p.c.)	(+) - ++	(+) - ++	(+) - +	φ - (+)	φ - (+)
lumina of glands (14 d p.c.)	+	+	+	φ - (+)	φ
stroma cells (12 d, 14 d p.c.)	φ	φ	+	+ - ++	++

3.2.3. Dissolution of the Blastocyst Coverings by Various Proteinases in vitro

14 μm unfixed cryostat sections from rabbit blastocysts at 7 d p.c. were subjected in vitro to various endo- and exopeptidases for 105 minutes at 38° C. In analyzing the results one must remember that there is no unit of activity which is valid for all proteinases, and that the activities cited are based on very different substrates and testing conditions. A comparison must, therefore, be made primarily on the basis of enzyme concentration (in mg/ml) which, due to the varying degrees of purity of the preparations, can only give us a rough foundation. In order to make more meaningful comparisons we determined the endopeptidase activity for all enzyme preparations using azocasein as substrate.

Table 10 gives a synopsis of the results obtained. It shows that *trypsin* effectively attacks the blastocyst coverings. The situation is similar for *papain,* although one must consider that the addition of cysteine which is necessary for this SH proteinase possibly has an auxiliary effect: treatment of control sections with 0.1 M cysteine in buffer caused a softening of the blastocyst coverings. This is not surprising in view of the assumption of a stabilization of the coverings through S—S bridges. The blastocyst coverings were also attacked by the endopeptidases *chymotrypsin, pepsin* and *elastase.* Under the conditions of the experiment no effect was seen with the endopeptidases *plasmin* or *collagenase* nor with any *exopeptidases* (see Tab. 10).

With incubation in liquid medium, such as was carried out in this investigation, the blasolemmase is largely washed away (Denker, 1971c). It is therefore unlikely that the blastolemmase activity of the 7 d sections used influenced the results to any great extent.

Table 1o. Experiments on dissolution of blastocyst coverings in the rabbit by various exo- and endopeptidases in vitro (stage 7 d p.c., 14 μm cryostat sections, incubation period 1 3/4 h at 38°C)

Enzyme (mU^{ACas}/mg[a])	Conditions	Lowest concentration still producing lysis	
		mg/ml	mU^{ACas}/ml
Papain (4000)	0.05 M Tris buffer pH 8.0; 0.005 M cysteine; 0.002 M EDTA	0.01	40
trypsin (45000)	0.067 M phosphate buffer pH 7.6; 0.005 M CaCl₂	0.05	2250
chymotrypsin (8600)	0.05 M Tris buffer pH 8.0; 0.025 M CaCl₂	0.5	4300
elastase (4000)	0.05 M Tris buffer pH 8.8	1.0	4000
pepsin (10400)	0.01 or 0.1 M Na-acetate-HCl buffer pH 1.8	0.5	5200
collagenase (40)[b]	0.05 M Tris buffer pH 7.2; 0.1 M CaCl₂	ϕ (> 1.0)	ϕ (> 40)
plasmin (910)	0.067 M phosphate buffer pH 7.6; 0.02 M lysine · HCl	ϕ (> 0.5)	ϕ (> 450)
leucine aminopeptidase (0)[c]	0.05 M Tris buffer pH 8.5; 0.004 M MgCl₂	ϕ (> 1.0)	ϕ[c]
amino acid arylamidase (0)[c]	0.067 M phosphate buffer pH 7.2	ϕ (> 1.0)	ϕ[c]
carboxypeptidase A (0)	0.05 M Tris buffer pH 7.6; 0.1 M NaCl; 0.007 M CoCl₂	ϕ (> 1.0)	ϕ[c]
carboxypeptidase B (0)[c,d]	0.05 M Tris buffer pH 7.6; 0.1 M NaCl	ϕ (> 0.5)[d]	ϕ[c,d]

[a]Values rounded-off; for definition of unit see 2.4; [b]At the limits of detectability; [c]No measurable endopeptidase activity with azocasein; [d]After treatment of the enzyme with 0.1 M DFP;

3.3. Investigations of the Physiological Role and Regulation of Blastocyst Proteinase in the Rabbit

3.3.1. Dependence on Trophoblast or Endometrium

3.3.1.1. Experiments with Blastocyst Models

As early as at the first description of blastolemmase activity in the rabbit blastocyst it was already supposed that it was being produced by the trophoblast (Denker, 1969, 1971c, 1972). In favor of this viewpoint is its localization on the surface of the blastocyst, although it appeared entirely possible that the enzyme was being synthesized in the endometrium and extruded in the uterine secretion (Kirchner, 1972a, 1975).

In model experiments the attempt was made to establish arguments for or against either of these two hypotheses. We tried to *eliminate the trophoblast from the system as a possible source of enzyme.* In a first series of experiments models for blastocyst coverings without trophoblast were used: unfertilized eggs and morulae (each with zona pellucida and mucoprotein layer) on the one hand and pure mucoprotein layer models on the other. The latter were produced by introducing foreign bodies the size of an egg (Sephadex-G-100 micro beads) into the tube where a mucoprotein covering layer would be formed around them as around a normal egg. These models were introduced into the uterus of pregnant or pseudopregnant females and left there for varying periods of time (Denker, 1975; Denker and Hafez, 1975). The gelatin substrate film test showed that such models develop *no* proteinase activity in the uterus worthy of mention, in contrast to the high activity of the control blastocysts. Furthermore, the mucoprotein layer of the models showed no morphological signs of the beginning of lysis for as long as it was observed, i. e. up to 7 or 8 d p.c.

The above mentioned models were only the size of morulae, but not of implantation stage blastocysts. It seemed therefore important to test on models the size of blastocysts whether or not the high proteinase activity found at the blastocyst surface was founded on a local stimulation of the release of a proteinase from the endometrium. Appropriate experiments with agarose beads have been mentioned in connection with the investigations of arylamidases (see 3.2.2.2.1.); similar beads found use in the preliminary experiments with the in vivo application of proteinase inhibitors (see 2.1.). The agarose beads were surrounded in the uterus by a variously thick, sometimes very well-developed covering of acid mucosubstances (equivalent of Böving's gloiolemma). But in this case as well the substrate film test showed *no significant proteinase activity* either in the covering nor around the model. These results can be taken as a clear indication for the *important role of the trophoblast* which must either produce the *enzyme (proenzyme) or essential cofactors (activators).*

3.3.1.2. Observations on the Lysis of the Coverings and on Implantation in Cases of Dystopic Implantation

Observations on *inversely orientated blastocysts in the uterus* suggest a role for the trophoblast in the blastolemmase activity and in the dissolution of the coverings. Normally the blastocysts orient themselves in the uterus with the embryonic disc facing mesometrially and with the abembryonic trophoblast facing antimesometrially (see 3.1.1.). We are interested here in those blastocysts who begin lysis of the coverings and undergo implantation orientated inversely. In a normal, non-experimentally tampered-with pregnancy they are very rarely found. More common are they after *various experimental conditions* and in *pathological situations.* In addition to the three cases previously mentioned (Denker, 1974b) we report here 11 other cases (see 3.3.2.2.). Details regarding the laboratory animals and their embryos will be discussed in 3.3.2. In the same group we had 186 normally orientated blastocysts. In all cases either an injection of proteinase inhibitor or NaCl solution into the uterine lumen was carried out 6 to 6 1/2 d p.c., or agarose beads were planted in the uterus. It appeared that the *manipulation of the uterus* itself was the triggering factor and not the administration of inhibitor: 5 cases of inversely orientated blastocysts were found in the control uteri into which only NaCl had been injected, and 5 cases in uteri treated with inhibitor. On the other hand there are indications for a possible surface- (or foreign body) -effect: next to 2 inversely orientated blastocysts we found a fibrin clot (apparently caused by bleeding after the injection) onto which the trophoblast had apparently attached itself; one of the blastocysts lay directly next to an agarose bead which had been placed in the uterus.

It held true for all the cases we observed that the *dissolution of the blastocyst coverings* began in the *region of the abembryonic trophoblast* regardless of any abnormal orientation, i. e. *even if the latter did not lie together with the normal "partner", the antimesometrial endometrium*. The coverings remained intact over the embryonic disc, although they lay antimesometrially in such cases. We conclude from this that the abembryonic trophoblast plays an important role in the dissolution of blastocyst coverings in the rabbit.

A high proteinase activity was found at the abembryonic pole of these malorientated blastocysts (or between it and the mesometrial uterine epithelium). This was, however, not always higher (as described in the first 3 observations, see Denker, 1974b) than the equally high activity often found antimesometrially between embryonic disc (coverings) and uterine epithelium.

3.3.2. Inhibition of Blastocyst Proteinase (Blastolemmase) and of Implantation in the Rabbit Through the Application of Proteinase Inhibitors in vivo

With this series of experiments we attempted to test whether or not the trophoplast-dependent proteinase (blastolemmase) activitiy of the blastocysts plays truly as essential a role in initiating implantation as we supposed. The experiments cited under 3.2.2.2.2. with the in vitro inhibition of blastolemmase form the foundation for our spectrum of acceptable inhibitors.

Type of Inhibitors

Two inhibitors with few known toxic side effects were primarily chosen for the in-vivo application:

1. The *basic trypsin-kallikrein inhibitor from bovine organs* (Kunitz) aprotinin, Trasylol®, used frequently in clinical medicine (Vogel et al., 1966; Trautschold et al., 1966; Werle, 1969, 1972);
2. The very low-molecular weight, microbial inhibitor *antipain* (Umezawa, 1972; Wingender, 1974; Wingender et al., 1975).

Both inhibit blastolemmase strongly and antipain the entoderm proteinase as well.

A smaller number of experiments were carried out using *p-nitrophenyl-p'-guandinobenzoate* (NPGB) (Chase and Shaw, 1970), an inhibitor which has found great interest because of its strong inhibition of the acrosin of living sperm (Wendt et al., 1975b) and therefore as a model for potential contraceptives; one animal received *boar seminal plasma trypsin-acrosin inhibitor* (SSPI) (a mixture of fractions I and II, see Fritz et al., 1975a; Fritz et al., 1976b; Tschesche et al., 1976). Both inhibit the blastolemmase very effectively, although not the entoderm proteinase.

For comparison we used in a series of animals *ε-aminocaproic acid* (EACA) which belongs to the group of the plasminogen activation inhibitors used therapeutically (inhibits plasmin also, but to a much lesser extent) (Bang, 1971; Andersson et al., 1965; Witt, 1975; Umezawa, 1972). It shows no in vitro inhibition of blastolemmase.

Inspired by a series of preliminary experiments (see 2.1.) the inhibitors were applied in dissolved form by intrauterine injection at 6 1/2 d p.c. Concentrations and doses per uterus can be read from Tab. 11 and 12. The animals were sacrificed exactly 7 1/2, 8 1/2, 9 1/2 or 11 1/2 d p.c. The light microscopical and electron microscopical morphology of the blastocysts and of the implantation and resorption sites in the uterus was studied. The proteinase activity was determined with the histochemical gelatin substrate film test (see 2.3.).

Table 11. Intrauterine application of proteinase inhibitors in the rabbit: dosage

	concentration in solution injected			dose per uterus[a]		
	mg/ml	Mol/l	ImU/ml[b]	mg/Uterus	mMol/Uterus	ImU/Uterus[b]
Trasylol® (MW 6512)	1.0	$1.54 \cdot 10^{-4}$	3800	0.6	$9.22 \cdot 10^{-5}$	2300
	7.0	$1.08 \cdot 10^{-3}$	27000	4.2	$6.46 \cdot 10^{-4}$	16200
	10.0	$1.54 \cdot 10^{-3}$	38600	6.0	$9.22 \cdot 10^{-4}$	23200
	20.0	$3.08 \cdot 10^{-3}$	77200	12.0	$1.85 \cdot 10^{-3}$	46300
antipain (MW 605)	2.0	$3.31 \cdot 10^{-3}$	24000	1.2	$1.98 \cdot 10^{-3}$	14400
	10.0	$1.65 \cdot 10^{-2}$	120000	6.0	$9.92 \cdot 10^{-3}$	72000
	20.0	$3.31 \cdot 10^{-2}$	240000	12.0	$1.98 \cdot 10^{-2}$	144000
SSPI I + II (MW 7000– 13000)	10.0	$\sim 1 \cdot 10^{-3}$	54000	6.0	$\sim 6 \cdot 10^{-4}$	32400
NPGB (MW 337)	0.1	$2.97 \cdot 10^{-4}$		0.06	$1.78 \cdot 10^{-4}$	
	0.33	$9.79 \cdot 10^{-4}$		0.2	$5.88 \cdot 10^{-4}$	
	0.5	$1.48 \cdot 10^{-3}$		0.3	$8.90 \cdot 10^{-4}$	
	1.0	$2.97 \cdot 10^{-3}$		0.6	$1.78 \cdot 10^{-3}$	
EACA (MW 131)	1.3	$9.92 \cdot 10^{-3}$		0.78	$5.95 \cdot 10^{-3}$	
	10.0	$7.63 \cdot 10^{-2}$		6.0	$4.58 \cdot 10^{-2}$	
	20.0	$1.53 \cdot 10^{-1}$		12.0	$9.16 \cdot 10^{-2}$	

[a] 3 x 0.2 = 0.6 ml was applied per uterus
[b] Trasylol®:   3800 ImU/mg
antipain: 12000 ImU/mg
SSPI I + II: 5400 ImU/mg

All values have been rounded-off

3.3.2.1. Morphology and Proteinase Histochemistry

The most important results are summarized in Table 12.

The *control injections* of NaCl in the uteri (or 10 % dimethyl formamide dissolved in NaCl solution) had no effect in any of the cases on implantation and development: the morphology as well as the proteinase activity did not vary from that of the untreated animals (compare Denker, 1970a and b, 1971c), apart from a certain frequency of cases with dystopic implantation (see 3.3.1.2. and 3.3.2.2.).

Treatment with Trasylol®

Trasylol® injected at a dosage of 0.6–12 mg/uterus 6 1/2 d p.c. causes a *very effective blockage of implantation* in the rabbit. The blastocyst coverings lysis is disturbed, and the trophoblast is unable to penetrate the endometrium or (with some blastocysts) only able to attach belatedly and only at a few spots. 7 1/2 d p.c. , after the application of 4.2–12 mg/uterus, no blastolemmase activity is demonstrable in the histochemical substrate film test (Fig. 18b, c) and there is only a suggestion of activity after 8 1/2 d

Table 12. Influence of intrauterine in vivo application of proteinase inhibitors on implantation in the rabbit

Trasylol®

time of sacrifice (d p.c.) / dose (mg/uterus) / I = inhibitor side, C = control side	7½ d 0,6 I	7½ d 0,6 C	7½ d 4,2 I	7½ d 4,2 C	7½ d 6 I	7½ d 6 C	7½ d 12 I	7½ d 12 C	8½ d 6 I	8½ d 6 C	8½ d 12 I	8½ d 12 C	9½ d 6 I	9½ d 6 C	9½ d 12 I	9½ d 12 C	11½ d 6 I	11½ d 6 C	11½ d 12 I	11½ d 12 C
1. Number of animals	1		1		1		4		2		3		1		1		1		1	
2. Number of corpora lutea	6	4	6	3	5	2	16	16	12	5	14	14	5	3	2	6	4	3	5	2
3. total number of implantation sites	4	4	3	3	4	2	11	14	11	4	13	13	5	1	2	5	4	3	5	2
normal-sized blastocysts	4	4	2	2	4	2	6	13	11	4	13	12	5	1	2	5	3	3	4	
too small blastocysts				1			3	1			11	1					2			
collapsed/degenerated embryos				1			2				1									
resorption sites													3				2	4	2	4
4. total number of blastocysts investigated																				
morphologically	4	2	2	3	2	2	5	5	8	3	12	5	2	1	2	1	3	2	5	2
protein in blastocyst cavity	2	2	2				5	5	1	3	1	5	1	1		1	2	2		1
collapsed/degenerated											1									
lysis of coverings: phase-correct	2			1	2		2	5		3		5		1	2	1		2		1
partially inhibited						1														
completely inhibited, coverings ruptured	2	2	2		2				6	2	10	1	2				3			
completely inhibited, coverings intact							5				1									
attachment of the trophoblast																				
abembryonically: normal		2		2				4		3		5		1				2		2
spotty attachment	1		1		1			1	5	3	8		1	1	2		1	2		
completely inhibited	3		2		1		5		3		4	4	1	1	2ᵃ	1	2		2	
embryonically			2	1						3				1	1			2		2
5. blastolemmase activity																				
total number of blastocysts investigated	2	1	2	3		2	3	3	4	3	8	2	1	1	2	1	1	1	1	1
activity > normal		1																		
normal		1						3		3	1	2					1ᵇ	1ᵇ	1ᵇ	1ᵇ
decreased	2		2	2	2	1		3	4		1	2	1		1	1		1ᵇ		
not detectable											5					1				
trophoblastic knobs positive	2			1			3				2				1	1	2		2	

ᵃ beginning of attachment, ᵇ no activity, ᶜ traces, ᵈ only slightly decreased

(continued)

Antipain

	1,2		7 1/2 d				8 1/2 d				9 1/2 d				11 1/2 d	
dose (mg/uterus)	1,2		6		12		6		12		6		12		12	
I = inhibitor side, C = control side	I	C	I	C	I	C	I	C	I	C	I	C	I	C	I	C
1. Number of animals	1		2		2		2		3		2		1		2	
2. number of corpora lutea	5	3	10	5	11	9	8	8	9	14	10	6	6	5	11	8
3. total number of implantation sites	5	2	10	5	10	8	7	6	5	12	10	5	6	3	11	8
normal-sized blastocysts	5	2	10	5	5	8	7	5	5	11	1	5	5	2	4	6
too small blastocysts					5			1			8		1			
collapsed/degenerated embryos							7	1	5							
resorption sites										1	1			1	7	2
4. total number of blastocysts investigated morphologically	5	1	9	4	5	8	4^c	1	4	5	3	2	6	2	3	3
protein in blastocyst cavity			3	4	2	8	4^c	1	3	6	3	2	2^c	2	3	3
collapsed/degenerated								1					1			
lysis of coverings: phase-correct	4	1	5	1	1	7	2	1	3	6	1	2	2	2	2	3
partially inhibited	1				3	1	2				2		4		1	
completely inhibited, coverings ruptured									1							
completely inhibited, coverings intact			4		1											
attachment of the trophoblast abembryonically: normal	5	1	3	1	7	1	1	1		6	1	2		2		3
spotty attachment			3	1	1	1	3		3		2		4		3	
completely inhibited			6		4				1				2			
embryonically										6		2		2	2	3
5. blastolemmase activity total number of blastocysts investigated	5	1	5	1	3	2	2	1	4	4	2	1	4	1	2	3
activity > normal	2	1					1		2		1				1	1
normal	3		2	1		2	1	1	2	4	1	1	2	1		
decreased			2		3		1									
not detectable			3								1		2		1^b	1^b
trophoblastic knobs positive	3				3		2		2				2	2		

a beginning of attachment, b no activity, c traces, d only slightly decreased

51

Table 12 (continued)

NPGB, EACA, SSPI I + H

	NPGB 7½d 0,06 I	C	NPGB 7½d 0,6 I	C	NPGB 8½d 0,6 I	C	NPGB 9½d 0,6 I	C	EACA 0,78 I	C	EACA 7½d 6 I	C	EACA 12 I	C	SSPI 7½d 6 I	C
time of sacrifice (d p.c.); dose (mg/uterus); I = inhibitor side, C = control side																
1. number of animals	1		2		1		1		1		1		1		1	
2. number of corpora lutea	4	6	8	7	6	4	6	6	5	3	4	3	6	6	5	2
3. total number of implantation sites	4	3	7	7	2	4	6	6	5	3	4	3	6	5	5	2
normal-sized blastocysts	2	3	2	7		4	1	6	4	3	4	3	6	5	5	2
too small blastocysts	1		1						1							
collapsed/degenerated embryos	1		4		2		2							1		
resorption sites	1						3									
4. total number of blastocysts investigated																
morphologically	4	2	7	2	2	1	3	1	5	1	4	1	5	3	5	2
protein in blastocyst cavity	4	2	3	2		1	1[c]	1	4	1	4	1	5	3	5	2
collapsed/degenerated			4		2				1							
lysis of coverings: phase-correct	4	2	3	2		1	1	1	4	1	4	1	3	3	5	
partially inhibited													2			2
completely inhibited, coverings ruptured			4		2		2									
completely inhibited, coverings intact									1							
attachment of the trophoblast																
abembryonically: normal	3	1	2	2		1		1	4	1	4		2	3	5	
spotty attachment	1	1	1				1						3			
completely inhibited			4		2		2		1			1				
embryonically						1	1(?)	1								
5. blastolemmase activity																
total number of blastocysts investigated	3	1	7	2	2	1	1	1	5	1	4	1	3	1	5	2
activity > normal																
normal	3	1	2	2		1	1	1	4	1	4	1	2	1	5[d]	2
decreased			1						1				1			
not detectable			4		2											
trophoblastic knobs positive			1				1								1	

[a] beginning of attachment, [t] no activity, [c] traces, [d] only slightly decreased

52

p. c.; traces can be found in antimesometrial-abembryonic spots 9 1/2 and 11 1/2 d p.c., which are still, however, weaker than the remnants of activity found normally at these stages (see 3.2.2.2.2.). After the application of the smallest dose tested (*0.6 mg/ uterus*) the blastolemmase activity is at 7 1/2 d p.c. completely inhibited in the blastocyst coverings, on the surface of the trophoblast and in the areas around the blastocysts, and no lysis of the coverings takes place. Interestingly though, a proteinase activity can then be traced to the *trophoblastic knobs,* although no activity can be discovered histochemically here in the control blastocysts (rather only after electrophoretic separation).

At a dosage of *4.2–12 mg/uterus* we find the following picture in detail for each of the individual stages:

7 1/2 d p.c. as opposed to the controls, the treated blastocysts are still able to be *flushed out* of the uterus. They are still *entirely surrounded by the blastocyst coverings* upon the outside of which a remarkable amount of uterine secretory material has adhered, especially where there are impressions in the surface of the mucosa. Trophoblastic knobs are developed, although they have not penetrated the blastocyst coverings and hence no attachment to the endometrium has taken place. Only a suggestion of *proteinase activity* can be found in the substrate film test at the surface of the epithelial cells in the depths of some endometrial crypts. The Trasylol®-resistant proteinase of the entoderm and the connective tissue cells in the endometrium and myometrium is also demonstrable as in the controls. The *cytological details* visible in the electron microscope show *no indication of any toxic effect* caused by the treatment with the inhibitor. The *blastocyst coverings* display a clear 3-ply *lamination;* there are *no signs of a beginning lysis* (Figs. 19, 20). The microvilli of the trophoblast and uterine epithelium only reach as far as the coverings but do not penetrate them. This is totally opposite to the situation in the control blastocysts where the coverings are dissolved in places especially above the trophoblastic knobs, while the remnants left lying between them show no lamination, and the trophoblastic microvilli stick into them, surrounded by a narrow empty appearing space (see 3.1.1., Figs. 4–7).

In all cases the *embryonic disc* at least displays retarded development or even signs of degeneration (pycnosis) to varying degrees. The cells do not reach a normal height and the mesoderm is scanty. A series of blastocysts could still be flushed out in toto from the uterus (see Tab. 13) during harvest of the uterine flushings for the determination of inhibitor concentration (see 3.3.2.3.). The degree of differentiation of the embryonic disc can be especially well studied on such total specimens. The embryonic discs of all of the embryos displayed either a clear retardation in development (no development of the primitive streak) or signs of degeneration (blurring of the normal outlines, irregular cell density, pycnoses) (Fig. 21.).

8 1/2 d p.c. the blastocysts treated with Trasylol® are further expanded as compared to a day before, but are not as large as the controls. The swelling of the uterus which marks the blastocyst sites is usually greater than would be representative of the diameter of the blastocysts: *the blastocysts do not usually quite fill up the uterine cavity.* Thus we have a picture which is never seen in a normal pregnancy in the rabbit: that of a freely floating or only focally attached blastocyst in a quantity of uterine fluid (Fig. 22b, 25). This fluid is granular and protein-rich and shows under the electron microscope that it contains abundant cell detritus but also relatively intact leucocytes and other free cells as well as cell fragments with no nuclei. The *blastocyst coverings*

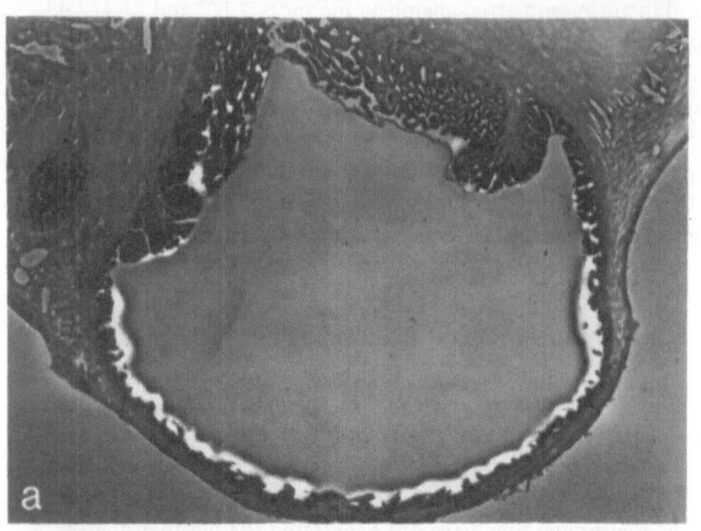

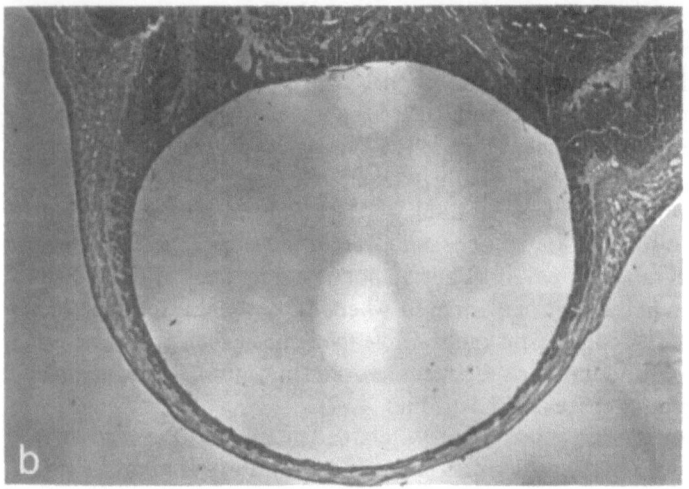

Fig. 18a–c. Rabbit, inhibition of implantation by the in vivo application of proteinase inhibitors. Trasylol®-application. Proteinase determination with the gelatin substrate film test.
(a) Control uterus; normal implantation site, 7 1/2 d p.c., cryostat section, substrate film test, incubation period 1 3/4 X 10. A high blastolemmase activity (bright lysis zone) can be detected in

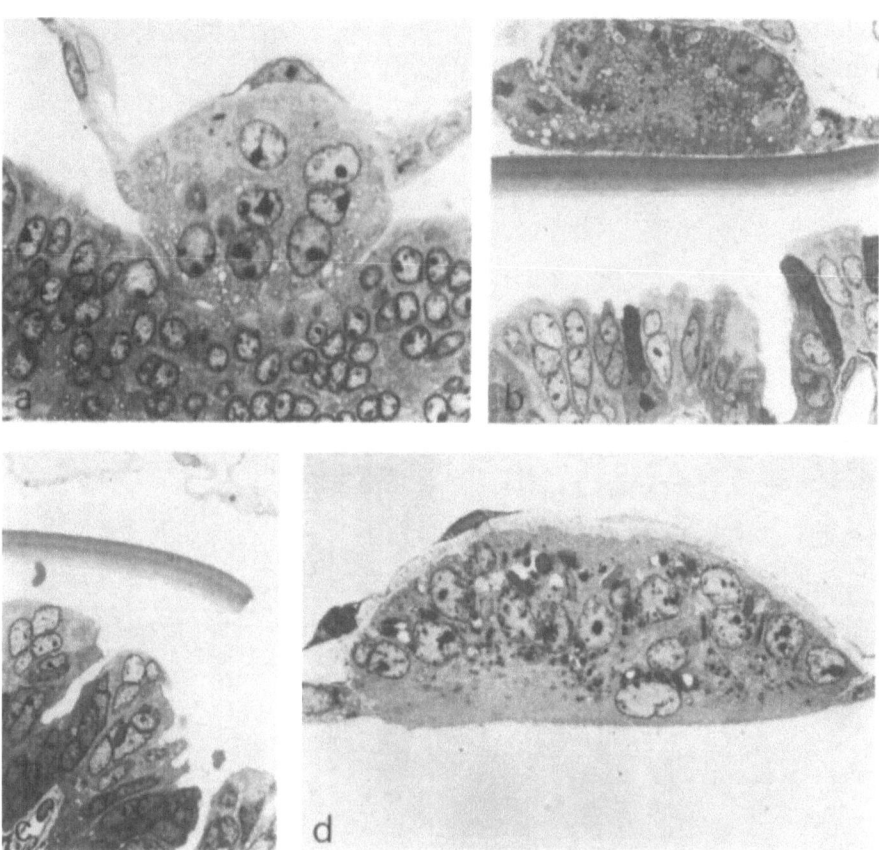

Fig. 19a–d. Rabbit, inhibition of implantation through the in vivo application of proteinase inhibitors. Trasylol® application. Morphology.
(a) Control uterus.Trophoblastic knob which has established contact with the uterine epithelium at several places. 7 1/2 d p.c., semi-thin section, toluidine blue, X 660. (b) Intrauterine application of 6 mg of Trasylol®. Trophoblast, blastocyst coverings and endometrium, 7 1/2 d p.c. Semi-thin section, toluidine blue, X 660. The blastocyst coverings are intact and show their lamination clearly. The trophoblastic knob has not established contact with the uterine epithelium. (c) Treatment as in b. In places the blastocyst coverings rupture because of the increasing expansion of the blastocysts. At the points of rupture no signs of lysis can be seen. (d) Same treatment as in Fig. b. Non-attached trophoblastic knob, 7 1/2 d p.c., semi-thin section, toluidine blue X 660. In the cytoplasm we see the crystalloid inclusions typical for the trophoblast. The surface facing the entoderm is covered by the characteristic protuberances (see Fig. 7)

the abembryonic (antimesometrial) hemisphere of the blastocyst, that is that region where the dissolution of the blastocyst coverings and the attachment of the trophoblastic knobs is already under way. In the region of the embryonic disc (above) the coverings remain intact. No significant proteinase reaction can be found there. (b) Intrauterine application of 12 mg Trasylol®. Blastocyst in uterus 7 1/2 d p.c., cryostat section, substrate film test, X 10 Blastolemmase completely inhibited, the blastocyst still surrounded by its coverings has preserved the typical round shape of the younger stage. It is tightly surrounded on all sides by endometrium, so that its walls cannot be clearly made out at this magnification. (c) As in b), application of 4.3 mg Trasylol®. Section of the abembryonic-antimesometrial region, X 200. The blastocyst covering whose dissolution is inhibited can be recognized as a dark band. Left in the picture a trophoblastic knob is shadily visible. Antimesometrial endometrium below. No detectable proteinase activity

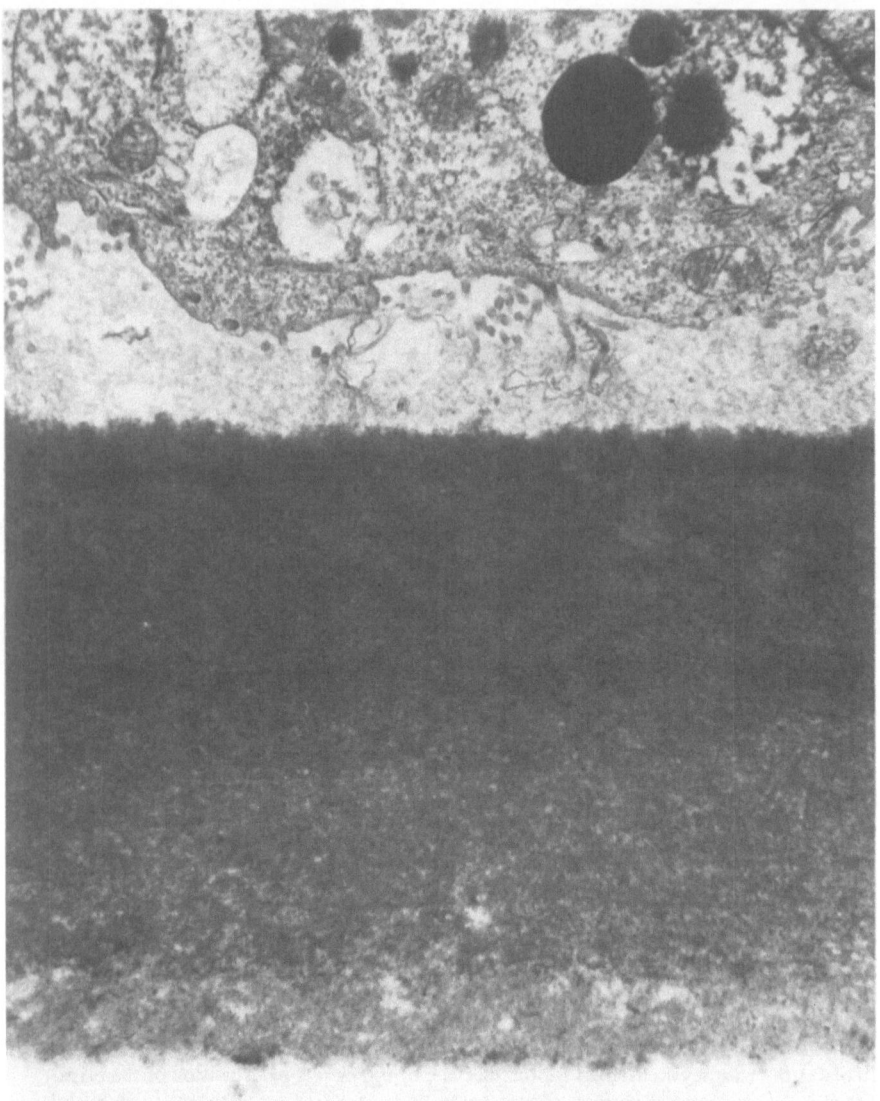

Fig. 20. Rabbit, inhibition of implantation through the in vivo application of proteinase inhibitors. Trasylol® application. Morphology. Intrauterine application of 6 mg Trasylol®. Trophoblast and blastocyst coverings, 7 1/2 d p.c., EM, X 10 000. The blastocyst coverings are composed of 3 clearly distinguishable layers, of which the middle one shows the concentric layering which is typical for the mucoprotein layer. The trophoblast is rich in vesicles filled with flocculous material (secretion?). Extracellularly, between the surface of the trophoblast and the blastocyst coverings copious amounts of a flocculous, fairly electron-dense substance

are still intact in some blastocysts and completely surround the embryo. In most blastocysts they are *ruptured,* and either remnants of them still lie against the surface of the blastocysts or larger fragments of them can be found throughout the uterine lumen. These fragments are quite well preserved and show *no clear signs of lysis,* but their edges are still well-defined (Fig. 23a, b). Where they still surrounded the blasto-cyst large quantities of thickened uterine secretion adhere in places over the outside

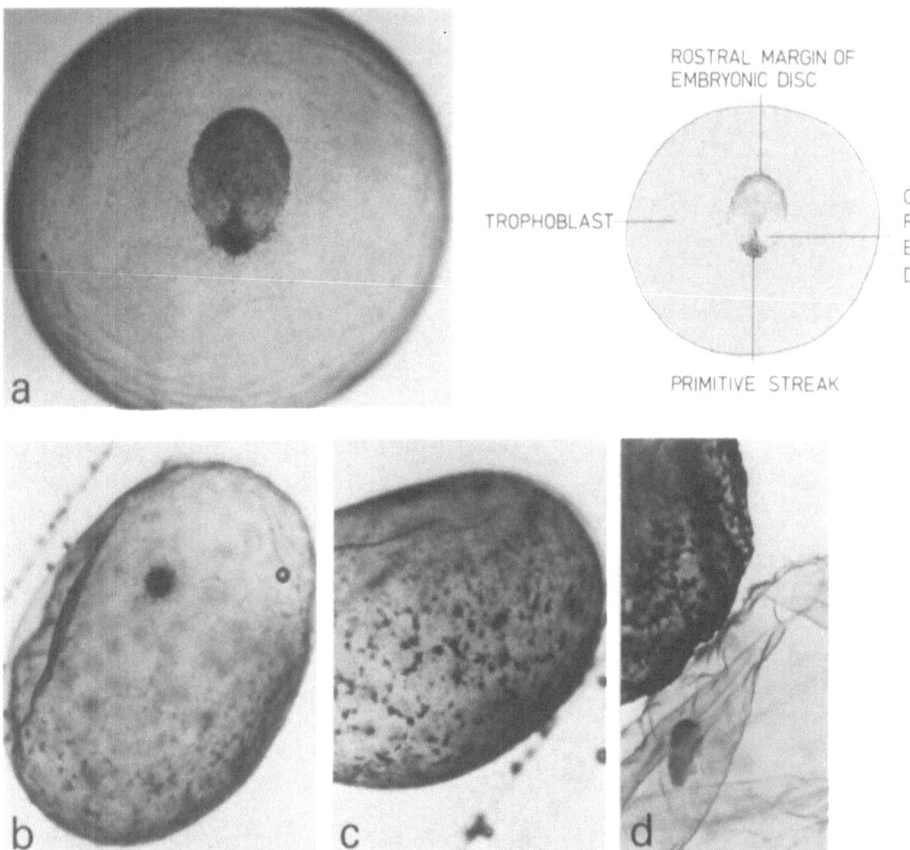

ROSTRAL MARGIN OF
EMBRYONIC DISC

TROPHOBLAST

CAUDAL
REGION OF
EMBRYONIC
DISC

PRIMITIVE STREAK

Fig. 21a–d. Rabbit, inhibition of implantation by the in vivo application of proteinase inhibitors.
Trasylol® application. Morphology of unsectioned blastocysts. Toluidine blue, X 16.
(a) Normal blastocyst flushed from control uterus, 6 1/2 d p.c. View of the embryonic disc.
The cell-rich rostral area is sharply demarcated from the trophoblast and can be distinguished
from a caudal area in which the cells are less densely aggregated and which is less well separated
from the trophoblast. The primitive streak seen here cannot be recognized in all of the blastocysts at this stage, but embryos should be expected to show at least a beginning differentiation
of the rostral and caudal areas at this time. (b) Blastocyst flushed from uterus at 7 1/2 d p.c.
after intrauterine application of 12 mg Trasylol®. The embryonic disc is degenerating. Blastocyst coverings still present. (c) Side-view of the blastocyst shown in (b). Trophoblastic knobs
(dark appearing structures) are developed in great numbers but no implantation has ocurred.
(d) As in (b), but blastocyst of another animal. Lateral regions of the trophoblast with numerous
trophoblastic knobs. The blastocyst coverings (lower right in picture) have ruptured and been
stripped off during flushing from the uterus, but are not yet lysed

surface. In other places an interesting, unusual sort of contact between the blastocysts
and the uterine epithelium forms: the blastocyst coverings lie in close contact to the
surface of the uterine epithelium, the microvilli have disappeared and *hemi-desmosome-
like structures* are formed which bind the surface of the uterine epithelial cells with
the coverings (Fig. 24).

The walls of the blastocysts are very thin an easily torn. The *trophoblast and entoderm cells* are stretched out thin. Trophoblastic knobs are varyingly well developed,

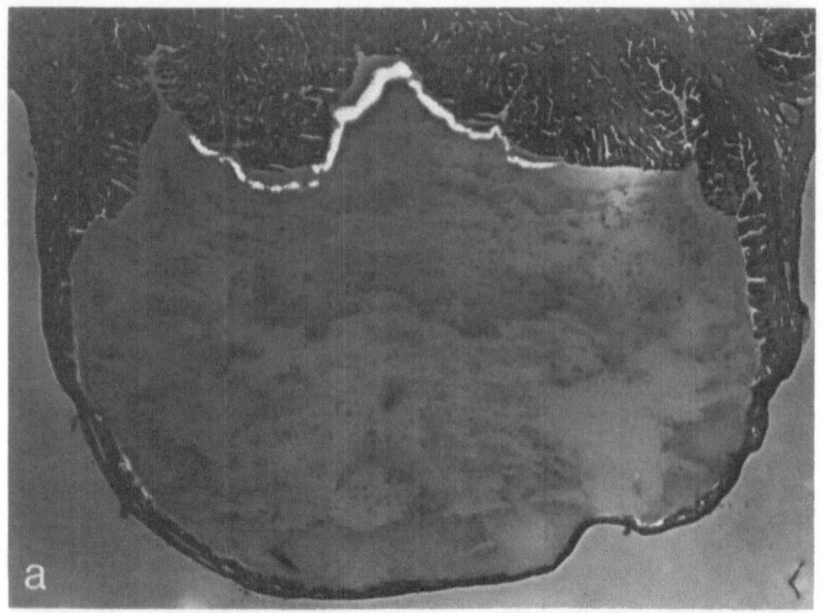

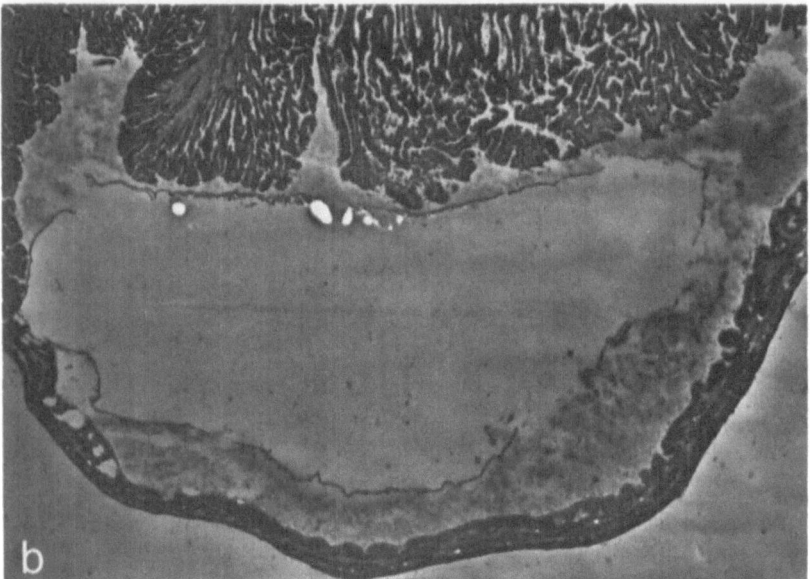

Fig. 22a and b

Fig. 22a–d. Rabbit, inhibition of implantation by the in vivo application of proteinase inhibitors. Trasylol[®] application. Proteinase determination with gelatin substrate film test.
(a) Control uterus; normal implantation site, 8 1/2 d p.c., cryostat section, substrate film test, incubation period 1 3/4 h, X 8.5. The expansion of the blastocyst has advanced further compared to the previous day (see Fig. 18a, note magnification). Quantities of maternal plasma proteins (cloudy precipitate) have passed into the yolk sac (former blastocyst cavity). The blastolemmase activity is reduced to a small remnant. At the embryonic pole a strong activity of a cathepsin-B-like enzyme can now be found (above), which can be distinguished from blastolemmase by in vitro inhibitor experiments. (b) Intrauterine application of 12 mg Trasylol[®]. Blastocyst in the uterus, 8 1/2 d p.c., cryostat section, substrate film test, incubation period 1 3/4 h, X 13.5. The blastocyst still lies un-

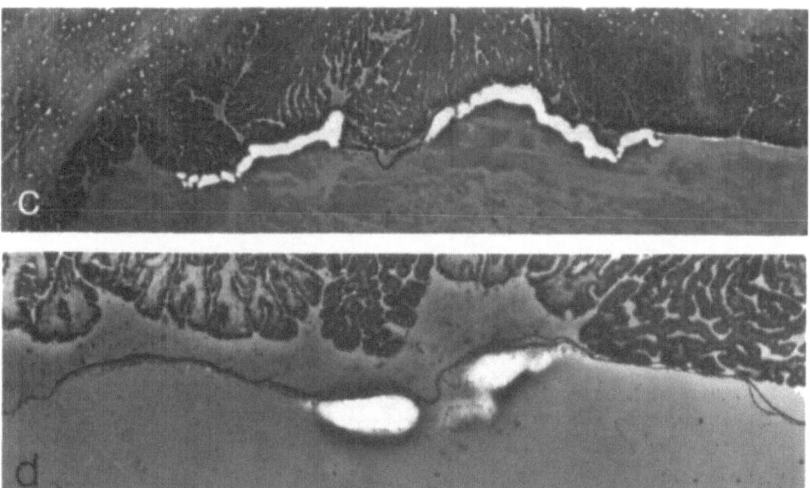

Fig. 22c and d

attached in the uterine cavity which is not quite filled-out. The blastocyst coverings are still largely intact (dark, band-like structures). Protein-rich liquid fills the uterine lumen around the blastocyst while the blastocyst cavity appears free of protein. In the region around the embryonic disc the cathepsin of the entoderm cells is detectable in spots, although the area taken by the positively reacting entoderm is abnormally small. Blastolemmase cannot be detected. (c) Control uterus as shown in a); mesometrial region of implantation site. X 12. The embryonic disc itself (in the center) shows no proteinase activity. Laterally from this, arranged symmetrically, is the cathepsin-rich entoderm (bright lysis zone). Above the latter lie those regions of the trophoblast which invade the mesometrial endometrium in the formation of the placenta. Blastolemmase cannot be detected here. Proteinase-positive stroma cells are disseminated in the endometrium; the reaction of these cells varies strongly from section to section. The blastocyst cavity is filled with heavily staining protein precipitates. (d) Intrauterine application of 6 mg Trasylol®. Blastocyst in uterus, 8 1/2 d p.c., embryonic disc region (mesometrial) as in c, X 23. The blastocyst still covered by the blastocyst coverings (dark line). Attachment to the mesometrial endometrium has not taken place. In the region of the hypoplastic embryonic disc, cathepsin activity can be detected in the under-developed entoderm. Blastocyst cavity practically free of protein in contrast to protein-rich (dark) uterine lumen

---

unusually small in some blastocysts, and displaying numerous pycnotic nuclei in others. They can also appear morphologically normal. The cells of the *abembryonic trophoblast* have all the typical cell organelles (see 3.1.1.) (including crystalloids and secretion granules of variously dense content). In some regions the number of granules even appears increased as compared to the controls. We find in addition similarly large, membrane-encased structures which contain vesicular inclusions and membranes.

In those blastocysts which have lost their coverings by rupture, we find a *tardy attachment of abembryonic trophoblast* on the uterine epithelium. It is often difficult to decide how intimate the contact actually is on cyrostat sections because of the extreme flatness of the trophoblastic cells. This must be taken into consideration in the evaluation of Table 12. A true *invasion* can be demonstrated well in some blasto-cysts on semi-thin sections (Fig. 23c) and under the electron microscope. In contrast to the controls, attachment does not take place in the entire abembryonic hemisphere of the blastocysts treated with inhibitor, but only in parts of it. No exact statements can be made about just how many blastocysts demonstrate this locally restricted at-tachment, because of the impossibility of producing serial thin sections of blastocysts

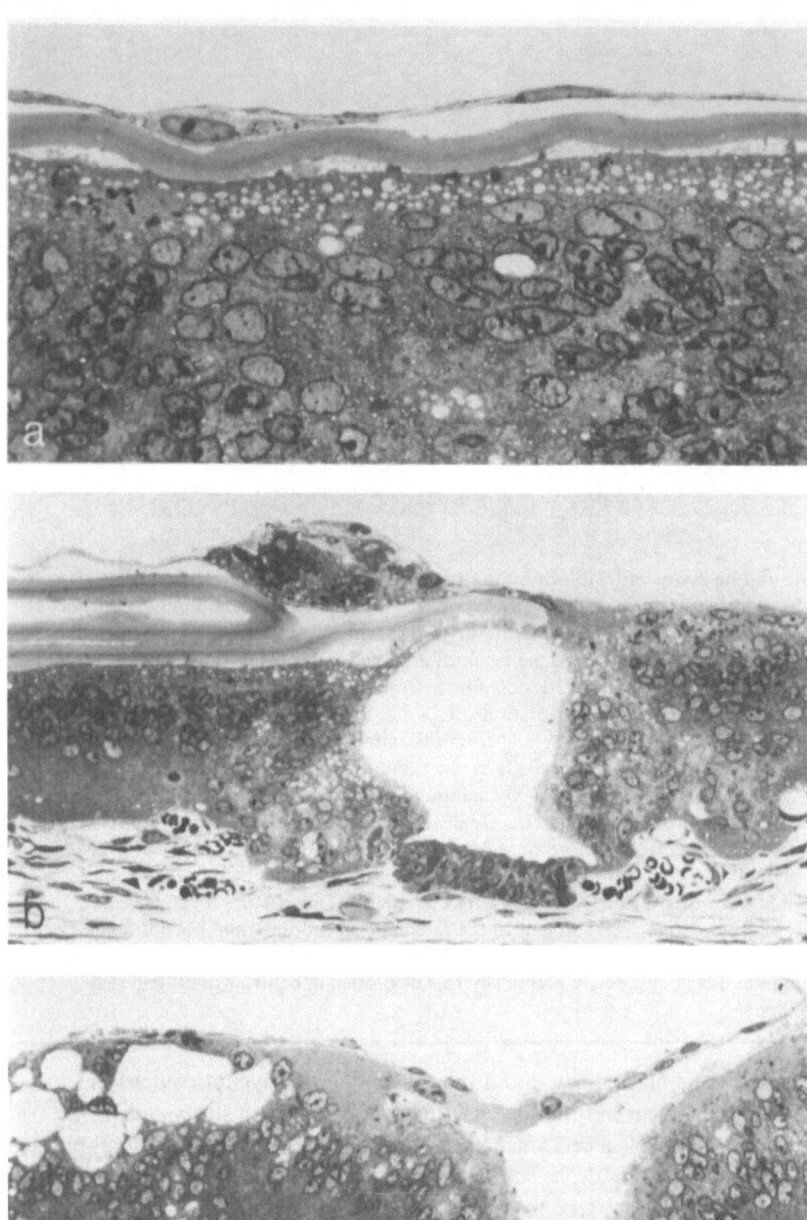

Fig. 23a–c. Rabbit, inhibition of implantation through the in vivo application of proteinase inhibitors. Trasylol® application. Morphology.
(a) Intrauterine application of 6 mg Trasylol®. Trophoblast, blastocyst coverings and endometrium of antimesometrial-abembryonic region, 8 1/2 d p.c., semi-thin section, toluidine blue, X 900. Even 2 days after application of the inhibitors and 1 day after the regular beginning of implantation the blastocyst coverings are still not dissolved. Their lamellar construction can still clearly be seen.

Fig. 24. Rabbit, inhibition of implantation through the in vivo application of proteinase inhibitors. Trasylol® application. Morphology. Intrauterine application of 6 mg of Trasylol®. Trophoblast, blastocyst coverings and uterine epithelial symplasm, 8 1/2 d p.c., EM, X 27,500. The trophoblast displays its typical basal lamina found on the side facing the blastocyst cavity. No entoderm is present in the abembryonic pole region represented here. The thickness of the blastocyst coverings is greatly reduced, though they are still relatively electron-dense. No lamination can be recognized in this region (as opposed to Fig. 23). Interestingly, half-desmosome-like structures form at the surface of the symplasmatically transformed uterine epithelium on the border with the blastocyst coverings lying closely attached. Apparently the blastocyst coverings are able to induce the formation of half-desmosomes just as the basal lamina is able to

Numerous vacuoles lie in the apical region of the symplasmatically transformed uterine epithelium. (b) Like a), X 350. The blastocyst coverings have ruptured and come to lie in the left half of the picture. At places where, after rupture of the coverings, the trophoblast comes into contact with the uterine epithelium, fusion can take place (right side of picture). The trophoblastic knob which still came to rest against the coverings is unable to establish contact with the endometrium because of the inhibition of the coverings lysis. The lumen of a crypt in the uterine epithelium is seen more towards the center. At its base the epithelium is still cellular. (c) Same blastocyst as in Fig. a and b, focal attachment of the trophoblast to the uterine epithelial symplasm, X 350. The trophoblast, which is also symplasmatically transformed (in the left hand side of the picture), is superficially fused with the uterine symplasm, but has apparently not reached the subepithelial capillary. At the border between both symplasms large, vacuole-like cavities form (left side of the picture)

of this size. The abembryonic trophoblast is *fused* in spots with the symplasm of the uterine cavum epithelium. Occasionally one finds remnants of the cell membranes at the interface between both symplasms, the uterine and that of the trophoblast. In the case of the uterine symplasm a multitude of lysosomal and lysophagosomal structures are especially noticeable. It is *not possible to confirm a topographical relationship between attached trophoblastic knobs and maternal subepithelial capillaries* like that of normal implantation, so that it is questionable whether or not this contact is fully functional with regard to the transport of metabolites. Bridges of cellular, flat trophoblast can be found between fusion sites. Gaps between themselves and the uterine epithelium are apparently widened by accumulated fluid in some places.

The *blastocyst cavity* appears *protein-free.* In the control blastocysts, however, protein could be demonstrated in the blastocyst lumen not only in fixed and embedded material where a flocculous precipitation is found, but also in unfixed material which was mounted on gelatin substrate films. The optical emptiness of the blastocyst cavity in the embryos treated with inhibitor contrasts strongly with the protein-richness of the surrounding uterine secretion (Fig. 22b). Such massive collections of uterine secretion around the blastocysts do not occur in the controls. In the controls at 8 1/2 d p.c. the attachment of the trophoblast to the endometrium of the placental area has begun at the *embryonic pole.* In no case can this be observed at that time after treatment with inhibitor, but one day later a tardy attempt at implantation can be recognized here in some blastocysts. In all cases the *embryonic disc* shows signs of a retardation in development or degeneration, although a strong variation in degree is observed.

The *entoderm* is relatively well developed. It has large granules, fairly similar to those of the trophoblast, but the contents of which are often only little electron-dense and homogeneous. Some of the granules stain especially strongly with toluidine blue in semi-thin sections. The entoderm has a high *proteinase activity,* which usually shows up even more strongly in the substrate film test than is the case with control blastocysts (Fig. 22). Inhibitor tests and pH optimum (see 3.2.2.2.2. and 4.4.2.1.) show it to be the typical catheptic entoderm proteinase.

Blastolemmase cannot be detected with certainty either in the trophoblast nor on the surface of the blastocysts nor at any other histological site. A low unidentified proteinase activity can only be seen at the surface of the uterine epithelial cells, especially

Table 13. Degeneration of the embryonic disc after in vivo application of proteinase inhibitors. Observations on whole blastocysts flushed out of the uterus

	Time after application of inhibitors or control injection of NaCl solution at 6 1/2 d p.c.				
	0 h	1 h	8 h	24 h	48 h
controls	17/2/0	1/1/0	16/2/1	[a]	[a]
Trasylol[®]	3/0/0	2/1/0	0/0/1	0/1/3	
antipain	7/1/0		5/0/6	0/0/2	0/0/1

Shown here is the number of blastocysts displaying normal/retarded/degenerated embryonic disc.

[a]could not be flushed out

in the region of the symplasmatically transformed antimesometrial and lateral cavum epithelium. Smaller still is the activity between implantation sites where it is largely concentrated in the deeper parts of the crypts.

In the endometrial stroma, mesometrially and antimesometrially, we find the *stroma cells* with their typical proteinase activity. Their numbers appear, in comparison to the controls, unchanged or at the most slightly decreased.

Similar to the situation with the controls, the *cavum epithelium of the uterus* is transformed to wide *symplasms* in the *neighborhood of the blastocysts* antimesometrially and at the top of the placental folds, which reach into the upper parts of the crypts. At the bases of the crypts, and partially up to the surface of them, the lateral cell walls are largely intact and the cells are bi- to multinucleate. The surface of the uterine epithelium, especially the antimesometrial cavum epithelium, has a thick lawn of extremely long microvilli. Frequently, these microvilli cover thin protuberances of the cell surface forming bush-like structures.

At *9 1/2 d p.c.* the unfolding of the embryonic body is already well advanced in the *controls,* the exocoel is well developed and the amnion is closed. After completion of the abembryonic-antimesometrial implantation (obplacentation) the mesometrial implantation which leads to the formation of the definitive chorioallantoid placenta is fully underway. The trophoblast of the embryonic pole is fused widely with the symplasmatically transformed cavum epithelium of the placental folds, has largely replaced it and grows down into the depths of it (see 3.1.1.) (Fig. 9).

In the uterus *treated with Trasylol*[®], on the other hand, the picture has changed relatively little from the preceding day (Fig. 25, cf. 22b). An *incomplete attachment* to the *antimesometrial* endometrium occurs with most blastocysts, although always only focally. The *blastocyst coverings* are ruptured and stripped-off and their remnants can be found in different places in the uterine lumen. The *uterine cavity* is not filled out by the blastocyst at the site of implantation. In contast to the regions of the uterine lumen between the sites of implantation it contains abundant protein, cell detritus and dead cells. In the *blastocyst cavity,* in contrast, *protein* is demonstrabel only in traces in a few blastocysts, while in the controls the blastocyst fluid is very protein-rich in this stage. The *trophoblast* of the treated blastocysts is extremely thin and the trophoblastic knobs display pycnotic nuclei. In some blastocysts the trophoblastic knobs are large and contain aggregates of vacuoles and lacunae and in other cases they are slight and apparently degenerated. The trophoblast of the *embryonic pole* starts in some of the embryos a tardy attempt to attach to the uterine epithelium. The embryonic disc itself is largely degenerated which can be seen especially clearly in the ectoderm and mesoderm. The *entoderm* appears still relatively healthy in comparison to the two aforementioned germ layers although still underdeveloped in comparison to the controls. One finds strong activity of the typical entoderm proteinase at the embryonic disc pole although the enzymatically active area does not spread out as far laterally as in the controls. *Blastolemmase activity* is not demonstrable or only barely demonstrable in the embryonic tissues or on their surface. The uterine epithelium displays slight to fair proteinase activity especially in those regions far removed from the blastocysts, but it reacts only minimally at the implantation site as is also the case with the uterine secretion.

The mesometrial endometrium displays a transformation to a *deciduoma* near the blastocysts; despite no attachment of the embryo adventitial, decidual cells can be demonstrated clearly although they are found less frequently than in the controls. The

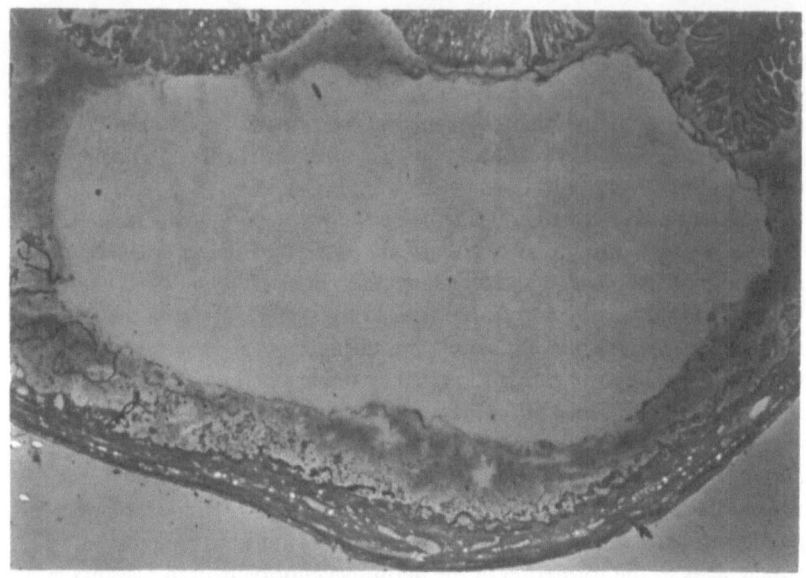

Fig. 25. Rabbit, inhibition of implantation by in vivo application of proteinase inhibitors. Trasylol[®] application. Proteinase determination with gelatin substrate film test. Intrauterine application of 6 mg Trasylol[®]. Blastocyst in uterus, 9 1/2 d p.c., cryostat section, substrate film test, incubation period 1 3/4 h, X 12. No attachment of blastocyst to endometrium can be seen in the section. The blastocyst coverings have been shed by rupture but their remains are not lysed and can be clearly recognized next to the blastocyst (in picture left bottom, right top). Embryonic disc not sectioned. The blastocyst cavity appears protein-free, but the uterine lumen protein-rich. The symplasms which arose from the antimesometrial uterine cavum epithelium are degenerating. Except for stroma cells, no proteinase activity is detectable

Fig. 26a–c. Rabbit, inhibition of implantation through in vivo application of proteinase inhibitors. Trasylol[®] application. (a) Extirpated left and right uterus, 11 1/2 d p.c. 6 mg Trasylol[®] had been injected into the right uterus. After inhibition of implantation one finds 4 resorption sites there, which only contain a deciduoma. At this stage no embryonic tissue can be detected. In the left uterus which only received a control injection of NaCl solution we find 3 sites of implantation with normal fetuses. (b) Control uterus; normal implantation site, 11 1/2 d p.c., cryostat section, proteinase determination with the gelatin substrate film test, incubation period 1 3/4 h, X 5.8. The embryo has been sectioned in the head region (eye cup, auditory vesicle). It is relatively closely surrounded by the amnion. It projects deeply into the wide yolk sac cavity (former blastocyst cavity), which is protein-rich. Above the embryo the exocoel has been sectioned; its contents are poor in protein. Its mesodermal wall is covered on the outside by extraembryonic entoderm, which shows typical cathepsin activity. The cathepsin-rich entoderm continues laterally. It has been sectioned again in the lower half of the picture because the sample was not cut exactly mesometrial-antimesometrially, but somewhat slanted. Blastolemmase activity, as in all stages after 7 1/2 d p.c., cannot be detected in significant amounts. We find only degenerated loose remnants of the anti-mesometrial uterine cavum epithelial symplasm (right and left in picture). (c) Intrauterine application of 6 mg Trasylol[®]. Resorption site. 11 1/2 d p.c., cryostat section, proteinase determination with the gelatin substrate film test as in b), X 11.5. The embryo which has been hindered in its implantation is completely degenerated; only detritus can be found in the uterine lumen. The deciduoma in the region of the placental fold is clearly recognizable. Proteinase activity is only found in individual stroma cells and in endometrial crypts further away from the sites of resorption. Note the different scale than in Fig. 26b

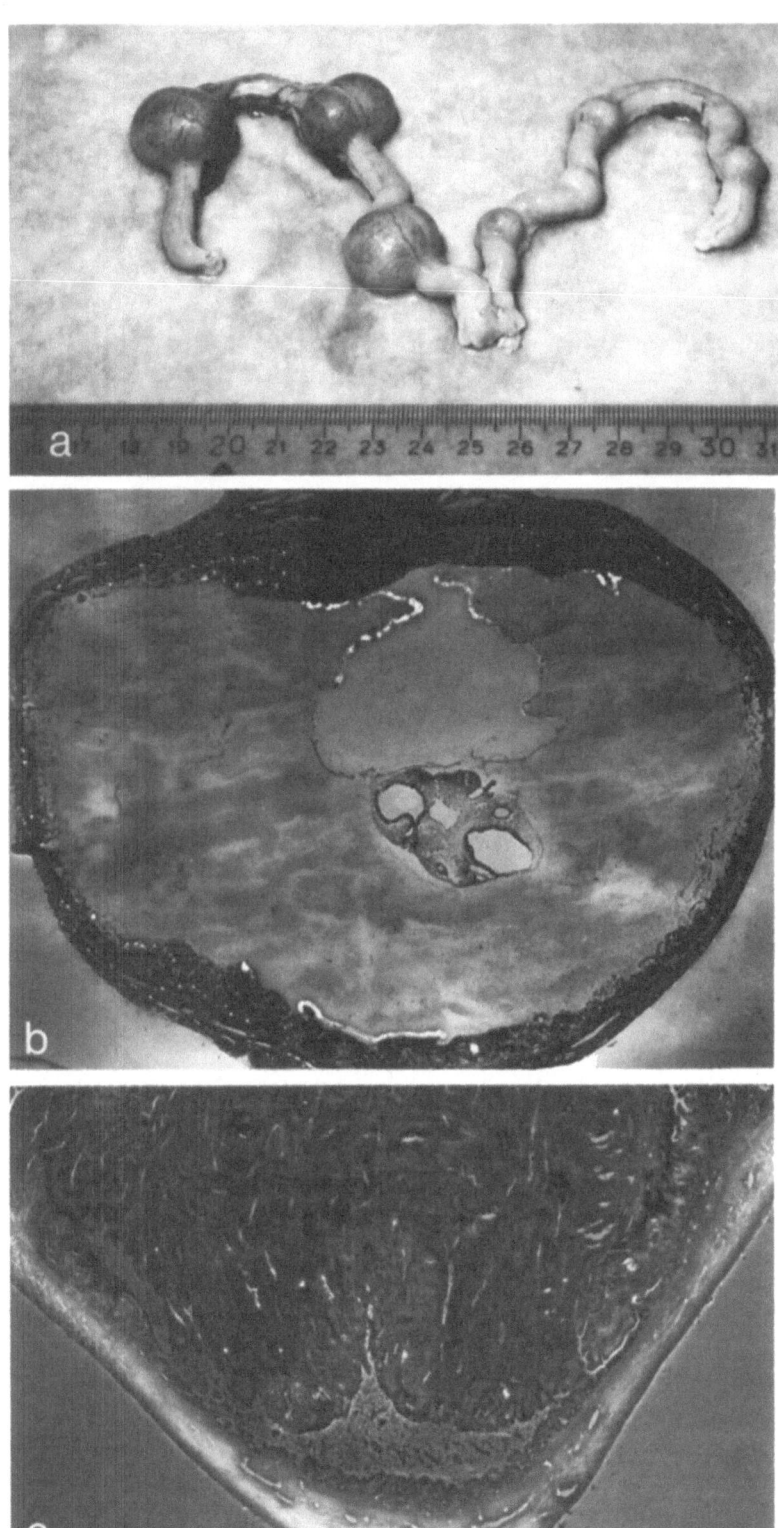

Fig. 26a–c

65

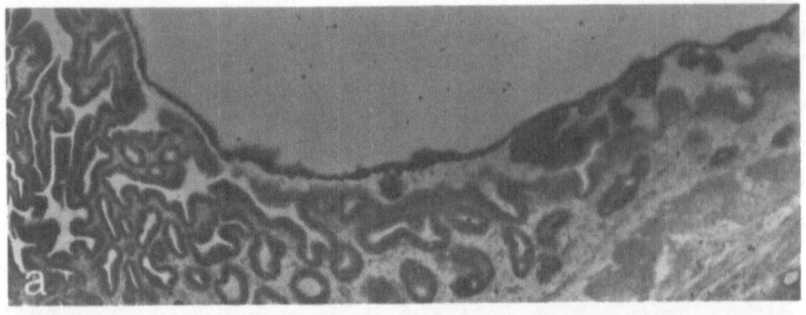

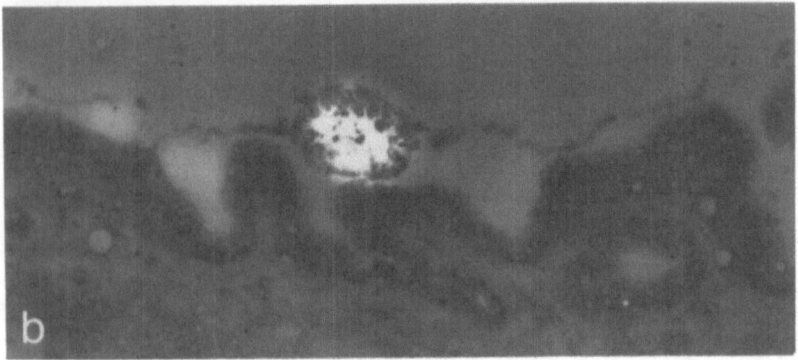

Fig. 27a and b. Rabbit, inhibition of implantation through in vivo application of proteinase inhibitors. Antipain application.
(a) Intrauterine application of 6 mg antipain. Section from the abembryonic trophoblast and from the antimesometrial endometrium, 7 1/2 d p.c., cryostat section, proteinase determination with gelatin substrate film test, incubation period 1 3/4 h, X 35. The blastocyst coverings are still intact (dark band). Very weak proteinase activity can only be detected in the endometrial crypts. Trophoblast and blastocyst coverings show no activity (normal control uterus: see Fig. 18a). (b) As Fig. a, but a dose of 1.2 mg X, 140. The trophoblastic knob displays high proteinase activity, but only traces are found between trophoblast and endometrium. Blastocyst coverings no longer existent

uterine epithelium is transformed to broad *symplasms* at the tops of the placental folds. Proteinase-positive stroma cells lie in the placental regions, but also in antimesometrial stroma and in mesometrial and antimesometrial myometrium. Their numbers appear to have slightly increased compared to the day before.

By *11 1/2 d p.c.* the formation of the placenta is far advanced in the *controls.* The embryo displays pharyngeal arches and limb buds. The exocoel is strongly expanded; the yolk sac is compressed antimesometrially, and its typical catheptic proteinase (see 3.2.2.2.2. and 4.4.2.1.) can be clearly demonstrated in the entoderm extending far in the lateral-antimesometrial direction (Fig. 26b).

In the *inhibitor-treated uteri,* on contrast, many embryos are undergoing *resorption;* here only cell detritus can be found remaining in the uterine lumen which, interestingly, shows hardly any proteinase activity. At such resorption sites the endometrium displays a well-developed *deciduoma* with numerous decidual cells and a fair number of basally lying proteinase-positive stroma cells (Fig. 26c). If the embryos have not yet undergone resorption they look in many respects like those of 9 1/2 d p.c.: one finds primarily a *trophoblast vesicle* which can be focally attached antimesome-

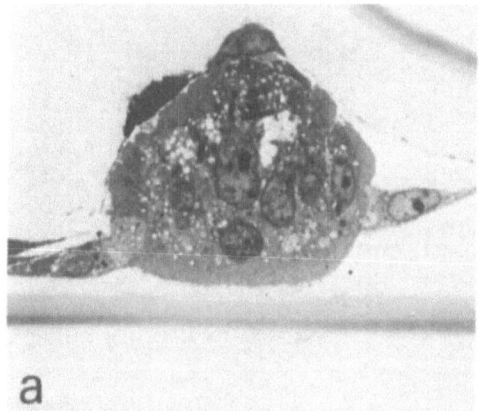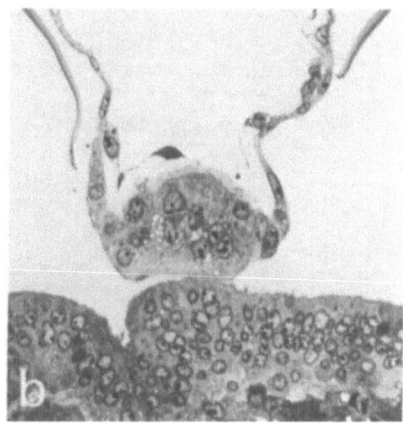

Fig. 28a and b. Rabbit, inhibition of implantation through in vivo application of proteinase inhibitors. Antipain application. Morphology.
(a) Intrauterine application of 6 mg antipain. Trophoblastic knob and blastocyst coverings, 7 1/2 d p.c., semi-thin section, toluidine blue, X 720. In contrast to the controls (see Fig. 5, 6, 19a) the blastocyst coverings are not yet dissolved and their lamination is still recognizable. One has the impression that the lysis has just begun in the region of the coverings lying against the trophoblastic knob. (b) As in a, X 280. Lysis of the blastocyst coverings can be successful where a trophoblastic knob lies. Such trophoblastic knobs are able to establish contact with the uterine epithelium. Laterally from this the coverings are still intact and show lamination (see, for comparison, normal control blastocysts, Figs. 5 and 6)

trially, contains little protein in its cavity and whose embryonic disc is largely degenerated, while the *entoderm* appears relatively well preserved and displays its typical, abundant proteinase activity at the embryonic pole. The uterine secretion shows proteinase activity between the embryos, which can hardly be found at the sites of implantation or resorption.

Antipain

If we take as criteria the coverings lysis (7 1/2 to 8 1/2 d p.c.) as well as the blastolemmase activity and the attachment of the trophoblastic knobs, then the inhibiting effect of antipain is less intense than that of Trasylol®. This also holds true when antipain is given in much higher doses (for example, 12 mg antipain as compared to 0.6 mg Trasylol®, which, because of the low molecular weight of antipain, is about 100 times the molar quantity, purity of sample taken into account). At 7 1/2 d p.c. the *blastolemmase activity* is reduced compared to the controls, but still clearly demonstrable. The *coverings lysis* is *delayed, but not completely inhibited:* the coverings show signs of beginning lysis 7 1/2 d p.c., and 8 1/2 d p.c. the blastocysts are always at least free of coverings in some spots, although larger remnants of the coverings are seen here than in the controls. The blastocysts appear to free themselves more through *lysis* than through mechanical splitting of the coverings (as in the case of Trasylol®). In semi-thin sections and under the electron microscope we see very clearly that the lysis of the coverings always *begins over the trophoblastic knobs:* laterally, well preserved thick remnants of the coverings can be found which still display their different layers (Fig. 28). In some blastocysts an extremely *thin lamella of blastocyst coverings*

67

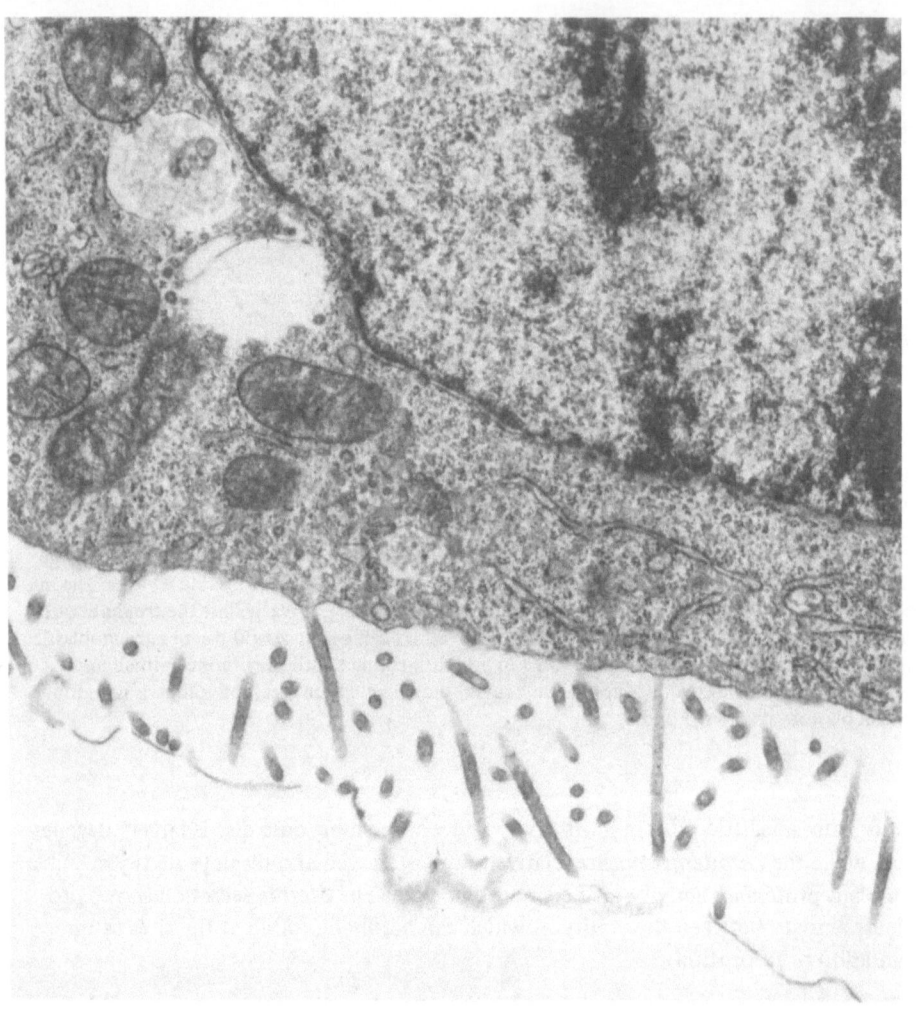

Fig. 29. Rabbit, inhibition of implantation through the in vivo application of proteinase inhibitors. Antipain application. Morphology. Intrauterine application of 6 mg antipain. Outer surface of the trophoblast, 7 1/2 d p.c., EM, X 16 900. An extremely thin lamella of blastocyst coverings material remains on the outer surface of the trophoblast on top of the long microvilli

material remains behind over wide areas of the blastocyst surface (Fig. 29). The microvilli of the surface of the trophoblast touch it, but do not penetrate it as in the untreated controls. Probably the thin remnant of the coverings cannot long stand the force of expansion and finally tears.

Nevertheless the antipain treatment seriously interferes with the formation of a functionally sufficient organ of metabolic exchange: the abembryonic/antimesometrial *attachment* of the trophoblast is *limited* to a small region of the abembryonic hemisphere. The *blastocyst* cavity remains almost *protein-free,* although a great deal of protein- and detritus-containing secretion collects in the neighborhood of the blastocyst in the uterine lumen, and as with Trasylol® a disproportion arises between the size of the uterine swelling and the diameter of the blastocyst. In many blastocysts the *embryonic disc* shows signs of degeneration already 8 hours after treatment (see Tab.

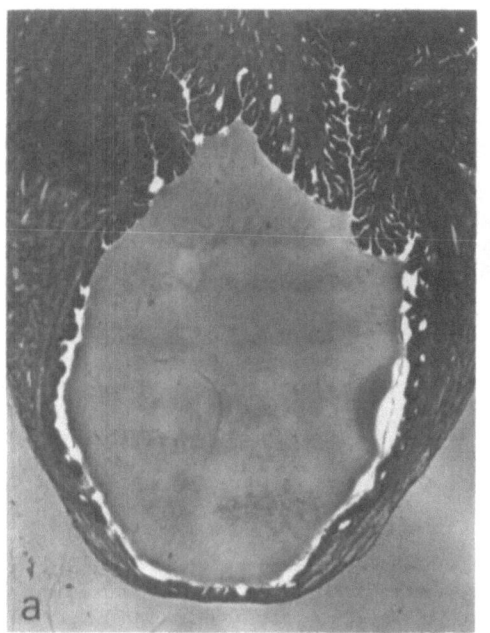
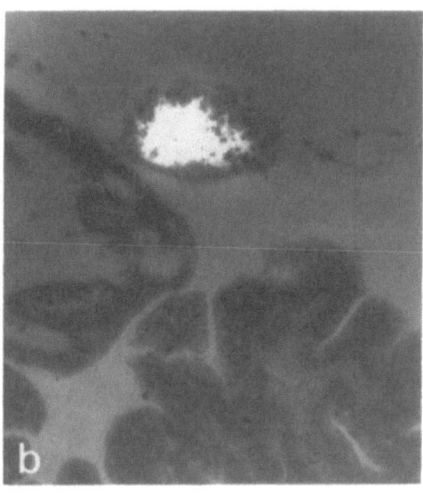

Fig. 30a and b. Rabbit, inhibition of implantation through the in vivo application of proteinase inhibitors. Antipain application. Proteinase determination with gelatin substrate film test. (a) Intrauterine application of 12 mg antipain. Blastocyst in the uterus 8 1/2 d p.c., cryostat section, substrate film test, incubation period 1 3/4 h, X 12. After decrease in the concentration of intrauterine inhibitor a delayed blastolemmase activity is found whose pattern of distribution is similar to that of the stage 7–7 1/2 d p.c. (see Fig. 18a) (normal 8 1/2 d p.c. implantation site see Fig. 22a). (b) As in Fig. a, 6 mg dose of antipain, section of trophoblast and endometrium, X 120. The unattached, large trophoblastic knob shows a high proteinase activity in its center which has lead to autolysis during incubation. In the remaining regions of the trophoblast, in the endometrium and uterine secretion no proteinase activity could be detected in this case

13). At first the *entoderm* remains fairly intact, and its typical proteinase activity can be demonstrated 7 1/2 d p.c. as well as in the following stages, although the enzyme is inhibited in vitro by antipain.

On the other hand certain phenomena can be observed after antipain treatment which one does not observe after the application of Trasylol®: the *uterine lumen* is distended by the streaming-in of secretion not only in the regions near the blastocysts, but in many cases between them as well. This is especially clear after 9 1/2 d p.c. It is often impossible to tell, in the intact uterus, which of the distentions contains blastocysts. Histological investigations show that not only do we have a collection of secretion at these blastocyst-free, widened areas, but also a *deciduoma formation* in various stages of development. The stroma of the endometrium is mostly hydropsically swollen in many places, which one rarely sees after treatment with Trasylol®. Particularly noticeable, especially 8 1/2 d p.c., is an increase in volume of those *trophoblastic knobs* which have not attached. These enlarged trophoblastic knobs show a considerable *proteinase activity* (Fig. 30b). This is especially noticeable because one can seldom directly demonstrate an activity of the trophoblast in the controls when tissue sections are studied using the histochemical substrate film test (see 3.2.2.2.2.). The

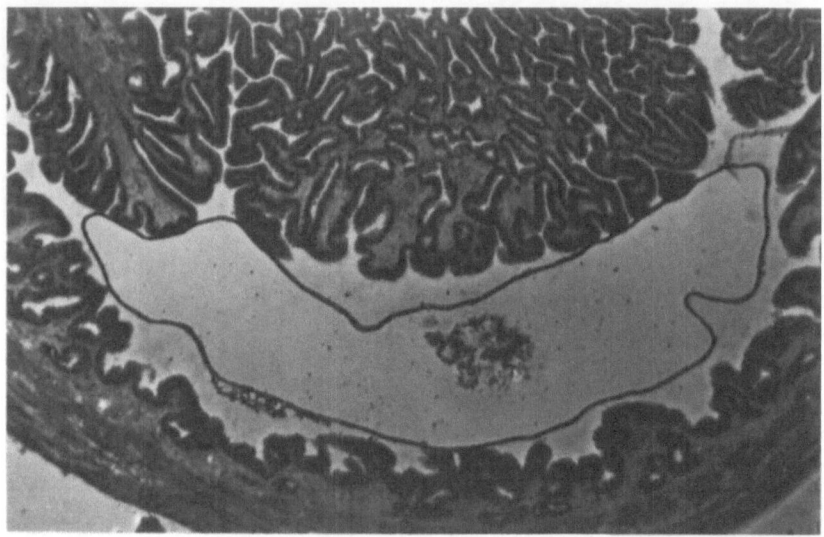

Fig. 31. Rabbit, inhibition of implantation through the in vivo application of proteinase inhibitors. NPGB application. Proteinase determination with the gelatin substrate film test. Intrauterine application of 0.6 mg NPGB. Blastocysts in uterus 7 1/2 d p.c., cryostat section, substrate film test, incubation period 1 3/4 h, X 35. The blastocyst coverings are completely intact (dark, band-like structures). Occasional mucus deposits on their outer surface. The blastocyst tissue is degenerated; only a group of dissociated, mostly degenerated cells can be found some of which still show proteinase activity (cathepsin) (entoderm cells?) Blastolemmase activity is not detectable (normal control see Fig. 18a)

enzyme activity originating in the trophoblastic knobs has basically the same spectrum of inhibitors and the same pH dependence as the blastolemmase. Possibly another proteinase takes part in the reaction as the inhibition by some blastolemmase inhibitors (SBTI, NPGB, ovomucoid) is incomplete and as a relatively strong activity remains demonstrable after lowering the pH to 5.0. There is a suggestion of activation by cysteine, although no inhibition by iodoacetamide could be shown.

Seminal Plasma Inhibitor (SSPI I + II)

Only one experiment with one animal was carried out using this inhibitor, as it appeared that no effective inhibiton of implantation could be achieved with relatively high doses (6 mg = 32400 ImU per uterus). 7 1/2 d p.c. the blastolemmase activity was weakened but still clearly demonstrable, the covering lysis was delayed relative to the controls, but not completely blocked.

NPGB

This inhibitor has a strong effect in low doses which varies, however, from blastocyst to blastocyst (see Tab. 12). A dose of 0.6 mg/uterus leads to a complete *inhibition of implantation* in many blastocysts. When the dosage was reduced to 0.06 mg/uterus, the blastolemmase activity, coverings lysis and implantation were normal in all blastocysts investigated; although one must keep in mind that only one animal was treated with this low dose.

70

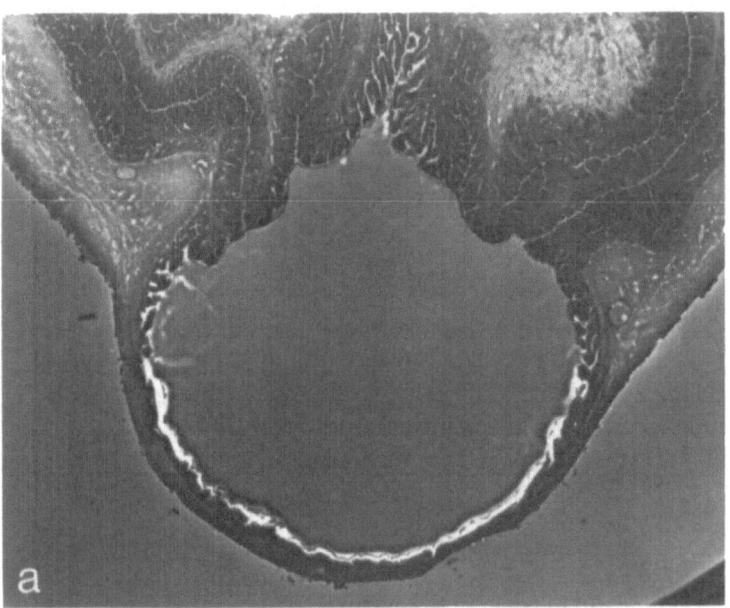

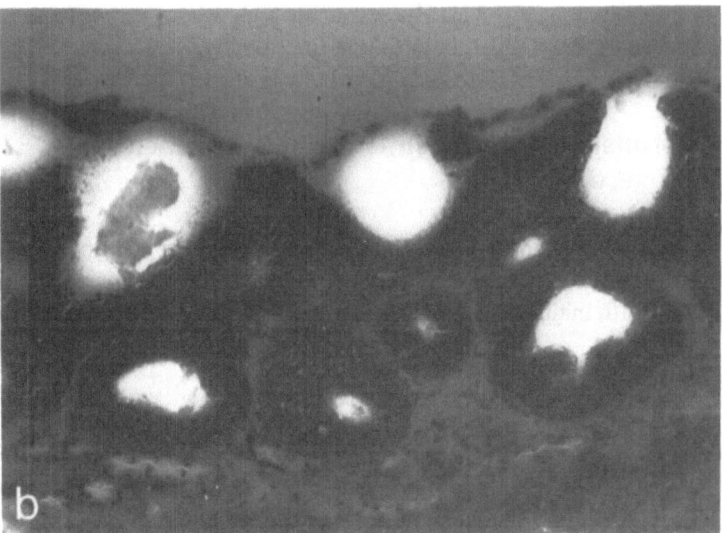

Fig. 32a and b. Rabbit, inhibition of implantation through in vivo application of proteinase inhibitors. EACA application. Proteinase determination with the gelatin substrate film test.
(a) Intrauterine application of 0.78 mg EACA. Blastocyst in uterus, 7 1/2 d p.c., cryostat section, substrate film test, incubation period 1 3/4 h, X 12. Blastolemmase activity, dissolution of the blastocyst coverings in abembryonic-antimesometrial region and attachment of the trophoblast are normal (see control Fig. 18a). The embryonic disc is pressed into a furrow in the mesometrial endometrium (above). The entoderm shows its typical cathepsin activity in that area. (b) As Fig. a, but dose 6 mg EACA. Section from the abembryonic trophoblast and the antimesometrial endometrium, X 140. The blastocyst coverings are lysed, a high blastolemmase activity can be found between trophoblast and uterine epithelium. Even with these high doses of EACA implantation begins in a normal fashion

The effects which can be seen in many blastocysts after application of 0.6 mg per uterus are very dramatic: complete inhibition of the blastolemmase activity (also when checked later), a lack of any signs of coverings lysis, but with, however, deposition of conspicuous amounts of mucus on the outside of the coverings (Fig. 31). The tissue of these blastocysts is apparently severely *damaged*, as opposed to the situation with the aforementioned inhibitors: 7 1/2 d p.c. it is already completely disintegrated, and inside the collapsed coverings only a clump of partly degenerated, partly still intact seeming cells remains. Certain of these cells show proteinase activity, which is possibly NPGB-resistent entoderm proteinase. 9 1/2 d p.c. these cells can no longer be found and have presumably died.

Interesting indeed is the fact that *other blastocysts* apparently survive the treatment *undamaged* (see Tab. 12). Already at 7 1/2 d p.c., one day after NPGB injection, one finds embryos which appear morphologically normal, whose coverings are undergoing lysis, who display a normal blastolemmase activity and whose trophoblast has begun invasion. Apparently the further development of these embryos is normal, at least as far as it was followed (to 9 1/2 d p.c.).

ε-Aminocaproic Acid (EACA)

Application of EACA in doses up to 6 mg/uterus has no effect on blastolemmase activity, coverings lysis or implantation (Fig. 32). The extremely high dose of 12 mg (ca. 10^{-1} mMol !)/uterus appears to show slight inhibition in some embryos (see Tab. 12).

3.3.2.2. Dytopic Implantation After the Intrauterine Application of Inhibitor or Control Injection

A noticeable accumulation of cases of the initiation of implantation with abnormal orientation of the blastocysts, which one seldom sees in untreated animals, was observed in the uteri treated with inhibitor as well as in the control uteri which had only

Table 14. Dystopic implantation of blastocysts after intrauterine injection of proteinase inhibitors or NaCl solution etc.

Stage (d p.c.)	Animal	Type of treatment	Number of inversely oriented blastocysts
7 1/2	1	control injection (NaCl)	1
	2	control injection (NaCl)	1
	3	control injection (NaCl)	2
	3	antipain injection (12 mg/uterus)	1
	4	agarose beads inserted	1
8	5	control injection (NaCl)	1
	5	antipain injection (1.2 mg/uterus)	1
8 1/2	6	Trasylol® injection (6 mg/uterus)	1
	7	antipain injection (12 mg/uterus)	1
9 1/2	8	antipain injection (6 mg/uterus)	1

received an injection of NaC1 (Tab. 14). Details of the morphology and proteinase histochemistry are described in the general discussion of dystopic implantation (see 3.3.1.2.).

3.3.2.3. Determination of the Time-dependent Decrease in the Concentration of the Inhibitors Injected into the Uterine Lumen

With these experiments we hoped to obtain preliminary data about how long after the intrauterine injection of inhibitor we can expect an actual effect to remain. Trasylol$^®$ and antipain were chosen in order to test whether or not the obove described unequally strong interference with implantation can be traced to the different speeds of elimination of these two inhibitors.

The results are given in Table 15 and Figs. 33 and 34. Every dot on the diagrams represents values from one rabbit. No statistical analysis was made because of the limited number of animals involved. One notices that the activity of both inhibitors decreases rapidly, in the uterine lumen, during the first 24 hrs. In the case of Trasylol$^®$ a remaining activity of about 10 % (average) can still be demonstrated 24 hrs. after injection, and at 48 hrs. the values are still clearly higher than those of the control side (see Tab. 15). 24 hrs. after application the activity of *antipain* lies within 1 % of its

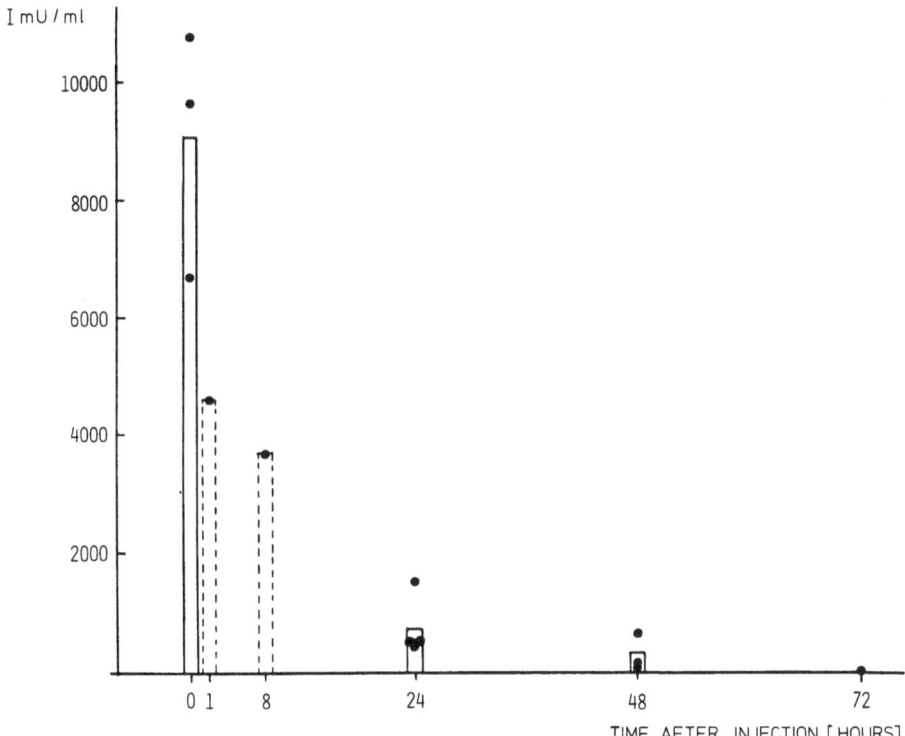

Fig. 33. Time-dependent decrease in the intrauterine Trasylol$^®$ concentration (expressed in ImU/ml uterine flushings) after a single injection into the uterine lumen at 6 1/2 d p.c. Inhibitor activity was measured against trypsin

Table 15. Time-dependent decrease in the concentration of intrauterally administered proteinase inhibitors in the uterine flushings of the rabbit

1. Trasylol®

Time after injection (hours)	Number of animals	Inhibitor side ImU/ml	ImU/µg prot.	Control side ImU/ml	ImU/µg prot.
0	3	3348	16.74	5.8	0.033
		5396.5	14.60	1.4	0.004
		4804	56.52	1.8	0.027
		x̄: 4516 (100%)	29.3 (100%)	3.0	0.021
1	1	2297 (50.9%)	12.76 (43.5%)	0	0
8	1	1846 (40.9%)	11.54 (39.4%)	4.7	0.037
24	4	751	3.95	1.8	0.047
		243.5	4.43	0	0
		239	5.98	1.9	0.004
		229	2.70	0	0
		x̄: 366 (8.1%)	4.27 (14.6%)	0.9	0.013
48	3	325	0.62	48.5	0.011
		52.5	2.63	0	0
		86.5	2.94		
		x̄: 155 (3.4%)	2.06 (7.0%)	24.3	0.006
72	1	20 (0.4%)	0.55 (1.9%)	1	0.048

2. Antipain

Time after injection (hours)	Number of animals	Inhibitor side ImU/ml	ImU/µg prot.	Control side ImU/ml	ImU/µg prot.
0	3	33333.3	128.20	16.7	0.07
		27506.7	171.92	13.6	0.085
		8295.5	24.40	22.2	0.25
		x̄: 23045.2 (100%)	108.20 (100%)	17.5	0.13
8	4	6129.0	22.70	0	0
		1186.8	13.96	0	0
		2239.0	15.99	1.65	0.02
		64.7	1.50	0	0
		x̄: 2404.9 (10.4%)	13.54 (12.5%)	0.4	0.005
24	4	69.1	0.53	0	0
		119.0	1.69	10.7	0.025
		83.4	2.77	0	0
		1.5	0.84	6.55	0.44
		x̄: 68.3 (0.3%)	1.46 (1.3%)	4.3	0.12
48	3	3,4	0.03	1.9	0.002
		33.1	0.08	0	0
		2.1	0.01	7.6	0.004
		x̄: 12.9 (0.06%)	0.04 (0.04%)	3.2	0.002

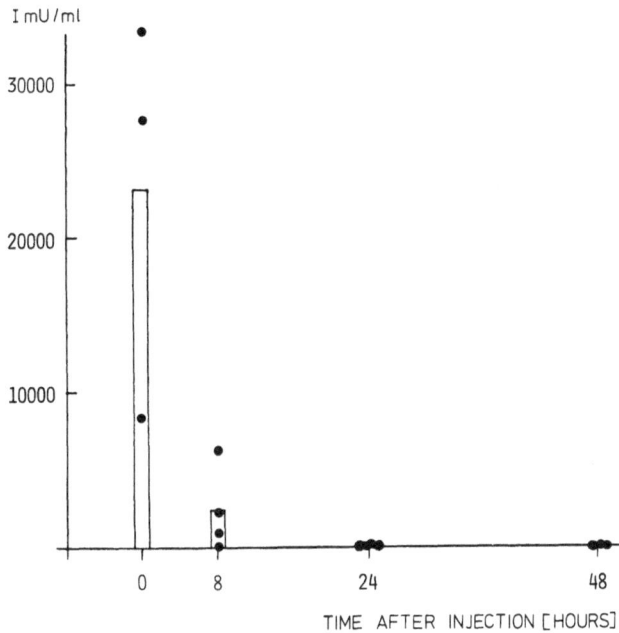

I mU/ml

TIME AFTER INJECTION [HOURS]

Fig. 34. Time-dependent decrease in the intrauterine antipain concentration (expressed in ImU/ml uterine flushings) after a single injection into the uterine lumen at 6 1/2 d p.c. Inhibitor activity was measured against trypsin

initial value and at 48 h can no longer be distinguished from the values for the control uteri. It is noteworthy that the values in the case of antipain vary especially strongly from animal to animal.

4. Discussion

The investigations here described regarding the morphology and histochemistry of the lysis of the blastocyst coverings in rabbits and cats lead us to suppose that *hydrolases* of the trophoblast or of the uterine epithelium or of both play a role here. At least in the case of the rabbit where the blastocyst coverings lysis is immediately followed by the establishment of contact between the trophoblast and uterine epithelium is it particularly seductive to ascribe a further role for the hydrolases in the establishment of cellular contact. This concept is partially supported by the inhibitor experiments described: the inhibition of proteolytic enzymes in vivo results in either complete blockage of the lysis of the coverings or at least a clear handicapping of it. The invasion of the endometrium is inhibited, although limited attachment will take place where a blastocyst manages, nevertheless, to free itself of its coverings.

Without doubt implantation is a very complex process which involves two organisms, mother and embryo, and therefore any simplified model of it, such as that of a simple lysis through secreted proteases must, a priori, appear questionable. We must therefore subject each individual step in the process of implantation to a more exact discussion.

4.1. Morphology of the Interaction Between Trophoblast and Endometrium in the Attachment Phase

Numerous electron microscopical studies of the past few years on a number of species have shown that there is a certain rule in the sequence of the various morphological stages during attachment of the trophoblast to the endometrium (for general description see Schlafke and Enders, 1975). In the first phase the central process is the establishment of an intimate cellular contact between trophoblast and uterine epithelium. During *apposition,* the free mobility of the blastocyst ceases and the parts of the endometrium which will enter into interaction with the trophoblast are determined. The mechanism of this apposition process and the morphological details vary strongly from species to species. For the immobilization of the embryo in species with small blastocysts (rodents) the local stromal oedema is important, whereas for animals with larger blastocysts (e.g. rabbit) the activity of the uterine musculature is important (Böving, 1959). In some species at least, the *blastocyst coverings* probably play a role in this process in that their adhesiveness increases (especially at the abembryonic pole of the blastocyst in the *rabbit*) shortly before dissolution begins. The rabbit trophoblast establishes contact with the uterine epithelium directly after the dissolution (see 3.1.1.). This close sequence of events has found great interest because of the possibility of a functional connection between the causal mechanisms. This might be valid also for other species (e. g. the ferret, see Enders and Schlafke, 1972). It apparently applies to murine mammals as well, at least in regular pregnancy (see Potts and Wilson, 1967; Smith and Wilson, 1974), although in these species the blastocysts are able to hatch mechanically from their zona and remain unattached as seen during delay of implantation (Psychoyos, 1966; Alloiteau and Psychoyos, 1966; Dickmann and De Feo, 1967; Orsini and McLaren, 1967; Dickmann, 1969; McLaren, 1970; Rumery and Blandau, 1971; Surani, 1975). In the cat (see 3.1.2.) and several other species (ungulates for example, see Bindon, 1971b; Leiser, 1975) the dissolution of the coverings and the attachment of the trophoblast are always clearly separated in time.

The situation in the human remains unknown as sections of the earliest attachment phase are lacking and the sections we do have of free human blastocysts have not been appropriately fixed for an investigation of the coverings (Böving, 1970; O'Rahilly, 1973). In the *guinea pig,* whose implantation is similar in many points to that of human embryos, extensions of the abembryonic trophoblast cells grow through the zona pellucida while dissolving it with proteolytic enzymes; directly afterwards they establish contact with the uterine epithelium (see Spee, 1901; Blandau, 1949, 1971; Parr, 1973).

After the blastocyst coverings have been disposed of, the trophoblast approaches the surface of the uterine cavum epithelium and the *microvilli of the trophoblast on twine loosely with the microvilli of the uterine epithelium.* Even at this stage of the apposition, the blastocysts can still be flushed out of the uterus using light pressure, although a certain degree of attachment has doubtless taken place. The morphologically observable intimacy of the contact and its tightness increase steadily thereafter.

In recent years interest has increased in the question which role the *glycoproteins of the cell surface* play in the establishment of this type of contact (see Denker, 1970a and b, Sartor, 1972; Enders and Schlafke, 1974; Schlafke and Enders, 1975). A thick layer of acid mucosubstances (MS) can be demonstrated using histochemical methods on the surface of the uterine epithelium. It is also present on the surface of the trophoblast (Bradbury et al., 1970; Holmes and Dickson, 1973; Martin et al., 1974; Nelson et al., 1976) although less predominant and perhaps histochemically different (Denker, 1970a). At least on the uterine epithelium it is considerably thicker than the glyco-calyx of other cells. Numerous investigations have yielded important evidence for the theory that cell surface carbohydrate substances are of great importance for the development of cell affinities and cell adhesion and therefore definitive in the determination of growth characteristics of cells, most obviously in the case of tumor cells and embryonic cells (see Curtis, 1967, 1970; Moscona, 1968, 1974; Lilien, 1969; Hause et al., 1970; Steinberg, 1970; Roth et al., 1971; Cook and Stoddart, 1973; Caravita and Zacchei, 1974; Lackie and Armstrong, 1975; Oppenheimer, 1975). It has been postulated that specific glycosyl-transferases mediate the selectivity of the contact (Roseman, 1970). It is possible that partial degradation of carbohydrate side chains by glycosidases which are demonstrable in uterine epithelium, uterine secretion and trophoblast (see 3.2.2.1.) influence the adhesiveness.

In terms of an electron microscopical classification the *adhesion stage* is reached when the surface membranes of the trophoblast and uterine epithelium not only approach one another focally, but also run parallel to one another over longer distances and are separated by a gap which is locally less than 200 Å wide. At first the intertwining of the microvilli on both surfaces becomes close and regular ("interdigitation"), but shortly thereafter the microvilli flatten and an irregular, wave-like contour remains of the membranes lying next to one other (Enders and Schlafke, 1967; Reinius, 1967; Bergström, 1971; Leiser, 1975; Parkening 1976). This sequence of events is found more or less typically in all species. In some species, however, (the rabbit for example) only some of the trophoblast cells establish the contact mentioned, whereas in other species (murine rodents) the adhesion takes place fairly uniformly over the entire trophoblast surface. In the latter species the opposed *uterine epithelial surfaces* undergo a similar hormonally controlled attachment reaction, so that the uterine lumen virtually disappears. This one does not observe in the rabbit (Nilsson, 1967; Hedlund et al., 1972).

The contact is finally reinforced by the development of cell junctions including desmosomes, which are regularly formed between trophoblast and uterine epithelium, although variously numerous (see Schlafke and Enders, 1975). This fact has found interest because of the possibility for a mediation of electrical coupling between cells by junctions (Bennett, 1973).

In the rabbit the final *penetration* of the uterine epithelium starts by a *fusion* of the trophoblastic knob with an epithelial cell, which can later include neighboring cells. After widening the space which these uterine cells originally occupied, a part of the trophoblastic cytoplasm including a few nuclei spreads in the direction of the subepithelial cappillaries (see 3.1.1.), breaks through the basal lamina and erodes the blood vessel (Fig. 4). Here the mixed syncytium appears to "take over" the cell walls of the uterine epithelial cells including the cell junctions. The fate of the uterine epithelial cell nuclei in this symplasm is not known exactly, but there is no evidence that they degenerate. On the other hand we find lysosomes and lysophagosome-like structures which in the following two days become very numerous in the mixed symplasm as well as in the regions of the forming uterine symplasm which are not fused with the trophoblast (Abraham et al., 1970). Whether or not they have a function in the sense of an "auto-digestion" of the uterine epithelium as has been postulated is at this time only a matter of speculation.

Schlafke and Enders (1975) compare the described penetration by fusion with *intrusion* (ferret), where the tongue-shaped projections of the trophoblast force themselves between the uterine epithelial cells, and with penetration by *displacement* (rat, mouse) where the uterine epithelium detaches over large areas from the basal membrane and degenerates. In any case one suspects that amoeboid movements of the trophoblast or its projections play a role; this supposition is logical in view of the density of microfilaments and microtubuli in its cytoplasm. In the rabbit (Larsen, 1961), rat and mouse (Potts, 1968; Tachi et al., 1970) invasion is halted for a while at the basal lamina. The mechanism of overcoming the basel lamina is unknown.

4.2. Morphological Observations on the Formation of the Blastocyst Coverings and Their Dissolution at Implantation

6 1/2 to 7 d p.c., i. e. before the beginning of lysis, the blastocyst coverings of the *rabbit* are composed of several layers (see 3.1.1.). The identification and origin of these layers are not completely clear (see Denker, 1970b). The innermost covering, the *zona pellucida*, is formed in the ovary. In its passage down the tubes the rabbit embryo receives another layer of deposited tubal secretion, the *mucoprotein layer*. These layers become both extremely thinned-out during the expansion of the blastocyst which reaches considerable proportions in this species. Böving has suggested (1954) that after the process of thinning-out the zona pellucida can simply no longer be seen with the light microscope. According to his concept, at the time of implantation the blastocyst coverings consist of the *mucoprotein layer* (*mucolemma*) and another layer derived from condensed uterine secretion, the so-called *gloiolemma*. Kirchner (1972b, 1973, 1975) also describes the blastocyst coverings as having two layers, but he supposes that the outer layer is the mucoprotein layer and the inner layer the zona pellucida; the latter having increased in volume by *swelling*. Such swelling must actually be assumed in other species where the blastocyst also undergoes great expansion, and where there is no evidence for the addition of extra material to the zona pellucida. This is the case in *carnivores* which, like the rabbit, have the central type of implantation (cat, see 3.1.2.; ferret, see Enders, 1971). The swelling must be caused by an increased hydratization due to chemical changes in the macromolecules of the coverings.

With regard to the composition of the rabbit blastocyst coverings, none of the concepts described above appear to mirror reality. As described above (3.1.1., 3.2.1.) we find, at 6–7 d p.c., in semi-thin sections and under the electron microscope not two, but at least three layers. Exhaustive substrate histochemical investigations have shown that the chemical *composition of the coverings changes between 5 and 7 d p.c.* (content of neuraminic acid, see 3.2.1., Denker, 1970a and b). A partial depolymerizing through the loss of sulfate ester-containing units which progresses from the inside to the outside has been postulated (Bacsich and Hamilton, 1954; Gothié, 1958, 1960; Denker, 1970b). Such changes in the macromolecules could be a reason for a change in degree of hydration. The thinned-out zona pellucida seems to be replaced, in the blastocyst stage, by a *new innermost layer secreted by the trophoblast* (Denker and Gerdes, in preparation). The mucoprotein layer takes the middle position in late blastocysts, while the outer layer represents Böving's gloiolemma. Uteroglobin, which possibly originated in the uterine secretion, can be demonstrated in the outer layer of blastocysts at 5 d p.c. with the light microscope using immunohistochemical methods; smaller amounts of it might penetrate the inner layers (Kirchner, 1972b). Unfortunately, the cited study did not include other stages.

Apart from the rabbit a lamination of the blastocyst coverings can be demonstrated in only a few *other placental mammals*, for example in the fur seal (Enders, 1971) where, however, the outer layer is identified as the zona pellucida and the inner layer as a special "subzonal layer". It appears entirely possible that the latter is equivalent to the innermost layer of rabbit blastocysts described above. Recently we have evidence from immunological investigations that also in those species with a morphologically single-layered zona, secretion material from the female genital tract seems to become incorporated (hamster: Fox and Shivers, 1975). Morphological evidence for the addition of secretory material on the surface of the zona has been found in the horse, the dog (for references see Bacsich and Hamiton, 1954), the mole (Heape, 1883), the badger, the cat (Amoroso, 1966) and the mink (Adams, 1973). The existence of a thin layer of uterine secretory material on the zona pellucida in the human is also being discussed (Böving, 1963).

In the rabbit the *dissolution of the coverings* begins in the abembryonic hemisphere usually in a ring-like zone around the abembryonic pole (see 3.1.1., Fig. 2, 3; Schoenfeld, 1903; Böving, 1954; Denker, 1970a; 1974b). The report by Kirchner (1972a) that it begins over the embryonic disc cannot be substantiated. The first sign of beginning lysis is always a *disappearance of lamination* (Fig. 5, 7); this appears typically always at those places where trophoblastic knobs lie against the coverings, while the lamination remains visible between the trophoblastic knobs. Evidence for an erosion exclusively from the outside to the inside, as Kirchner reported from investigations with unstained native sections under the interference contrast microscope, cannot be supported by the evidence of fixed and stained semi- and ultra-thin sections. Rather one finds, after partial inhibition of the lysis of the coverings by proteinase inhibitors, the thickness of the blastocyst coverings greatly reduced where a trophoblastic knob lies, and as in these blastocysts the layering of the coverings often remains well recognizable, it becomes clear that the *inner layer* often disappears first over the trophoblastic knobs (see Fig. 28a). With uninhibited or only partially inhibited lysis we see under the electron microscope that the microvilli of the surface of the trophoblast reach into the inner parts of the coverings (see 3.1.1., Fig. 7). They are surrounded by a narrow, electron-optically empty region, and it is difficult to decide whether this is an artifact of shrinkage or a region of lysis. It is conceivable that with the constant expansion of the blastocysts, free lysis regions could not appear between the surface of the trophoblast and the coverings as the trophoblast is tightly pressed against the still existent remnants of the coverings. In normal implantation, between the outer surface of the coverings and the uterine epithelium we find a fissure full of low molecular weight material which invites the interpretation that the dissolution of the coverings also begins simultaneously *from the outside* (uterine secretion).

These morphological investigations at least show clearly that we have here a lytic phenomenon and not a rupture of the coverings; which role the trophoblast and uterine epithelium play here cannot be determined on the basis of these investigations alone.

4.3. Concepts About the Chemical Composition of the Blastocyst Coverings and Possible Pathways for Their Enzymatic Degradation

Chemical analysis has not yet been carried out on the blastocyst coverings. Tests of this sort meet with the greatest methodological difficulties not only because of the small amounts of test material available, but because these highly polymer substances are difficult to get into solution without inducing chemical modifications (Denker and Petzoldt, unpublished; Hamner, personal communication). In the rabbit a composition of *high molecular carbohydrate substances* and *protein* could be demonstrated (see 3.2.1.; Denker, 1970a and b). The carbohydrate units carry *neuraminic acid* on their ends, probably largely in 0-acetylated form, which is indicated by the differing neuraminidase susceptibility before and after saponification (literature see 3.2.1.). Apart from the bound neuraminic acid, *sulfuric acid ester groups* determine the polyanionic character of these substances. All histochemical characteristics point to a relatedness with other epithelial mucins. The histochemist has to restrict himself basically to demonstrating the presence of functional groups (vicinal OH-groups, sulfuric acid ester and neuraminic acid groups) and their association with proteins. Further statements, especially those regarding the bonds possibly linking carbohydrate chains and proteins, can only be made after biochemical analysis. For this reason we are obligated to restrict ourselves to models which we have derived from our knowledge of biochemically researched epithelial mucins.

The *principal structure of an epithelial glycoprotein* is exemplified in Fig. 35; details of this structural outline are to be regarded as an arbitrarily chosen example (Gottschalk, 1964, 1966; Kent, 1967; Tappel, 1969). Carbohydrate side chains are bound through O-seryl-(or N-aspartyl-) bonds to a protein backbone. They can be of varying length and have varying monosaccaride composition. Typically, neuraminic acid is found at the ends (esterified as sialic acids). Various OH-groups are candidates for esterification with sulfuric acid; the position of sulfate ester groups in epithelial mucins is not exactly known.

The biological degradation of such macromolecules is carried out by *glycosidases, peptidases* (endopeptidases, exopeptidases: amino- or carboxypeptidases) and enzymes which cleave the protein-carbohydrate bond (for example: O-seryl-N-acetylgalactosaminide glycosidase, Bhargava et al., 1966, Buddecke et al., 1969; aspartamido-N-acetylglucosamine amidohydrolase, Conchie and Strachan, 1969) (see Fig. 35). It is known from numerous biochemical investigations that the negatively charged groups (neuraminic acid, sulfuric acid ester) and, to a certain extent, uncharged carbohydrate side chains lend the mucosubstances a relative *resistence to degradation by proteinases,* although to different degrees for different proteinases (Faillard and Pribilla, 1964; Gottschalk, 1964; Gottschalk and Fazekas de St. Groth, 1960; Martin et al., 1967). For the in vivo degradation in lysosomes, for example, a certain sequence therefore has been assumed: first degradation of the carbohydrate side chains by neuraminidase and successive degradation by the glycosidases specific for the individual sugars, then further degradation by endo- and exopeptidases (see Mahadevan et al., 1969; Tappel, 1969).

4.4. On the Physiological Role of Glycosidases and Proteases in the Dissolution of the Blastocyst Coverings and in Implantation

4.4.1. Enzymatic Dissolution of the Blastocyst Coverings in vitro and Correlation with the Patterns of Distribution of the Corresponding Enzymes

Actual experimental evidence was found that the above mentioned theories may also hold for the dissolution of the blastocyst coverings in the rabbit: the terminal sialic acids also lend relative resistence here to degradation by proteinases, and their removal by neuraminidase insures that the proteolytic cleavage by trypsin is facilitated (see

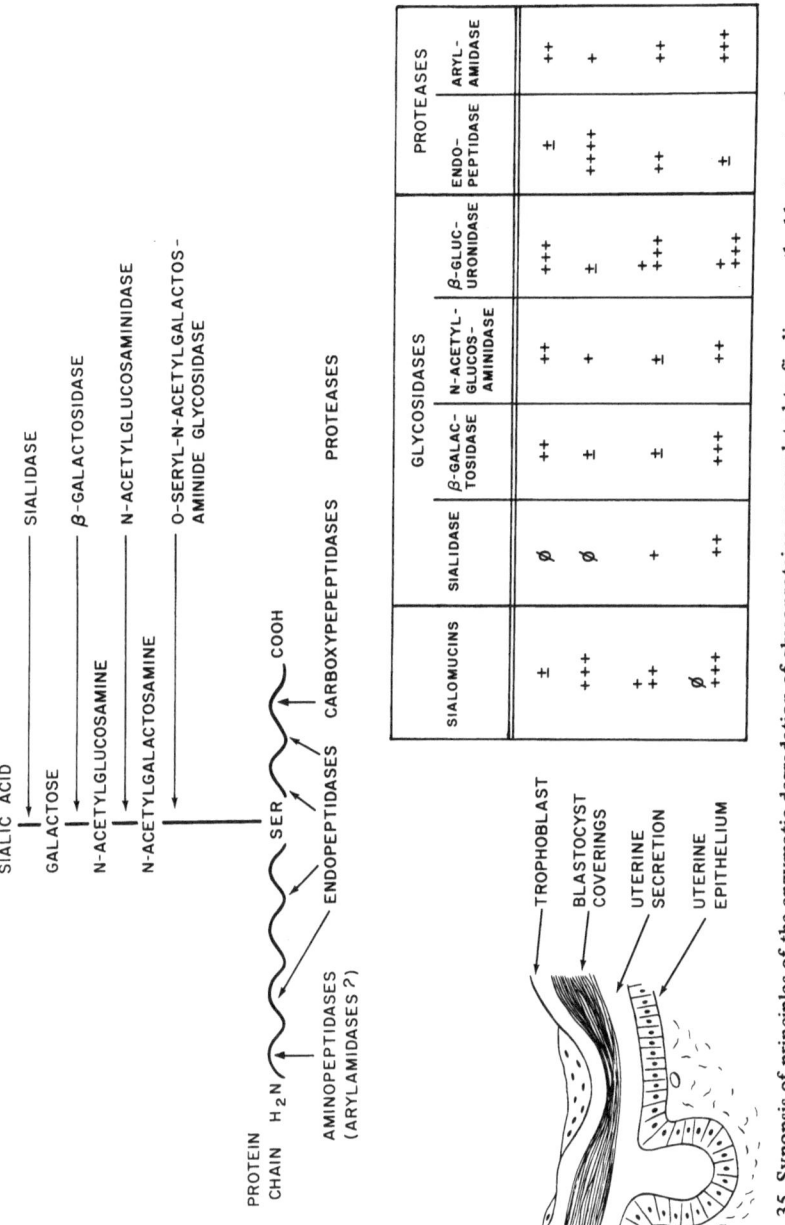

Fig. 35. Synopsis of principles of the enzymatic degradation of glycoproteins as correlated to findings on the blastocyst and uterus. In the upper part of the drawing the basic composition of a glycoprotein (mucosubstance) from a protein backbone and carbohydrate side chains (arbitrarily chosen example) is sketched along with the enzymes which are possible candidates for hydrolysis of the individual bonds. In the lower part the pattern of distribution of such neuraminic acid-rich mucosubstances (sialomucins) and their corresponding glycosidases and proteases which are found at the interface between mother and embryo during the stages of implantation in the rabbit are shown

3.2.1.). Neuraminidase by itself, however, is not capable of lysing the blastocyst coverings; it can only operate supportively. The effect of incubating with other glycosidases in vitro was investigated in rodents and it was found that is was not sufficient to lyse the coverings (Bowman and McLaren, 1970). Exopeptidases are also incapable of producing complete lysis; the action of endopeptidases is necessary (proteinases, see below).

Glycosidases which can operate supportively or preparatively on the blastocyst coverings lysis as mentioned above can be found in the trophoblast, in the uterine epithelium and uterine secretion, and some of them show a rise in activity around the time of implantation (see 3.2.2.1., Fig. 35). β-Galactosidase, N-acetylglucosaminidase and β-glucuronidase can be found in the uterine epithelium and in the trophoblast abembryonic-antimesometrially in the region where the coverings lysis begins. A preliminary attempt to localize neuraminidase, however, showed maximal activity not in those structures which take part in the lysis of the blastocyst coverings, but rather in the mesometrial endometrium in the paraplacental folds where the remnants of the coverings of the embryonic hemisphere finally come to rest after they are pushed away from the embryo, having become freely movable after the dissolution of the abembryonic part (see p. 35; Denker, 1971b).

In other species, glycosidases have also been demonstrated in the *endometrium and uterine secretion;* characteristic changes in the activity during the preimplantation phase take place here as well (β-glucuronidase: Fishman and Fishman, 1944; Conchie and Findlay, 1959; Ringler, 1961; Thiery and Willighagen, 1963; Manning et al., 1966; Wood and Psychoyos, 1967; Woessner, 1969; Anton et al., 1969; Wood et al., 1969; Wood and Barley, 1970; Abraham et al., 1970; Moulton, 1974; Linford and Iosson, 1975. Glucosaminidase (hexosaminidase): Conchie and Findlay, 1959; Coleman et al., 1967; Anton et al., 1969; Linford and Iosson, 1975. β-Galactosidase: Conchie and Findlay, 1959; Woessner, 1969. Neuraminidase: Ganguly et al., 1976). However, we have no specific investigation of the effects of the inhibition of glycosidases in vivo on the lysis of the coverings and on implantation. In discussing the possible physiological role of the glycosidases mentioned, we have to caution that they must be primarily considered as lysosomal enzymes, which have their pH optimum in the acid range. The uterine milieu is alkaline. Since the pH optimum depends on the substrate and conditions of the reaction, and as long as no specific experiments have been carried out with isolated uterine glycosidases and natural substrates, the question must remain whether these enzymes may actually play a part in the lysis of the blastocyst coverings in the uterine lumen or not.

Like the glycosidases, *exopeptidases* alone are unable to lyse the blastocyst coverings in vitro (see 3.2.3., Tab. 10). These enzymes may, however, have a supportive function as aminopeptidases show noticeably high activity in the uterine epithelium and uterine secretion. Particularly interesting is the *arylamidase I* (see 3.2.2.2.1.) demonstrable with leucine-β-napthylamide which has been shown to have an especially high activity in rabbit uterine secretion. The enzyme was referred to as leucine aminopeptidase at first (Denker, 1969, 1971c, Petry et al., 1970; Beier et al., 1971, 1972a). More comprehensive investigations (see 3.2.2.2.1.; van Hoorn, unpublished; van Hoorn and Denker, 1975) have shown, however, that classical leucine aminopeptidase (EC 3.4.11.1) (pH optimum between 9 and 10, cleavage of leucineamide and leucine hydrazide) cannot be found in measurable amounts in rabbit uterine secretion; the enzyme found there is more likely to be an amino acid arylamidase because of its substrate specificity (preference for amino acid arylamides) and its other biochemical properties such as pH optimum (about 8.0) and activation by Co^{++} (see Rehfeld and Schultka, 1967; Rehfeld et al., 1967; Pearse, 1972; Bergmeyer, 1974; Tamura et al., 1975). The enzyme is certainly not related to cathepsin B (as discussed by Sylvén, 1968, Sylvén and Snellman, 1968; Snellman, 1969; Otto and Riesenkoenig, 1975) because of the inhibiting effect of cysteine and EDTA and because of a lack of inhibition by iodoacetamide. Amino acid arylamidases have not been classified but are more likely related to EC 3.4.11.2 than EC 3.4.11.1).

A special function for arylamidase I in the physiology of the later preimplantation phase has been suspected because of its noticeably high activity in the uterine secretion at this time (Denker, 1969, 1971c; Beier et al., 1971, 1972a). It is especially interesting that the activity in the uterine epithelium appears to be influenced by the blastocyst immediately before implantation starts (see 3.2.2.2.1.; Denker and van Hoorn, 1974; van Hoorn and Denker, 1975; Denker, 1976c): the enzymatic activity in uterine epithelial cells declines faster and earlier in the neighborhood of the blastocyst than in the regions further away from the blastocyst or than in the pseudopregnant endometrium (Fig. 12). Apparently the *arylamidase extrusion of the surrounding uterine epithelium is increased by some stimulus released by the blastocyst whose chemical nature is still unknown.* Alternatively, it is thinkable that an inhibition of the enzyme synthesis (with the turnover rate remaining constant) could be an explanation for the fall in activity. In favor of the stimulation of secretion theory is a comparison of the activities in the uterine flushings of normally pregnant uteri (with blastocysts) and pseudopregnant uteri (without blastocysts): the peak of activity in the uterine secretion is reached earlier in the presence of blastocysts (Fig. 11). It is interesting that the enzyme can also be found in blastocyst tissue (trophoblast), so that we could discuss whether or not it is possibly taken up by the blastocyst from the uterine secretion. In the *cat* we find clear activity of a very similar (if not identical) enzyme in the trophoblast, although the uterine epithelium and secretion show no reaction; this probably indicates a synthesis by the trophoblast itself.

The role of the enzyme in the physiological processes at this phase of pregnancy is as yet unknown. The fall in activity in the uterine secretion prior to 7 d p.c. speaks against a participation in the dissolution of the blastocyst coverings and the initiation of implantation. The enzyme could play a role in the supplying of amino acids which are taken up by and incorporated into the blastocyst during this phase (for an outline of the literature see Jaszczak and Hafez 1972; Biggers and Stern, 1973; Petzoldt et al., 1973). Furthermore we must also consider a participation in the activation or inactivation of biologically active uterine peptides similar to that which has been shown for the inactivation of releasing hormones by arylamidases (Kuhl and Taubert, 1975a and b).

It is interesting that, in rough comparison to the blastocysts, *copper IUDs* also cause a dramatic fall in the arylamidase activity in the surrounding uterine epithelium and secretion in the rabbit, caused in this case, however, by a direct inhibition of the enzyme (see 3.2.2.2.1.; Denker, 1976d; Denker and Kühnel, 1977). It remains to be seen what significance this possibly has in the mechanism behind the efficacy of copper IUDs, which is still unexplained at a molecular level (see Duncan and Wheeler, 1975). Investigations of this sort could very well prove relevant in medicine, as the *human* endometrium, uterine secretion and placenta (as well as those of *other species*) contain large quantities of aminopeptidases and other peptidases (Hanson and Smith, 1948; Smith, 1948; Semm, 1958; Fuhrmann, 1959; Albers et al., 1961; Hopsu et al., 1961; Lammes, 1963; Vollrath, 1965; Beckman et al., 1966, James, 1966; Schmidt et al., 1966; Christie, 1967; Zsolnai et al., 1967, Seelig and Roemheld, 1969; Joshi et al., 1970; Walter et al., 1971; Bergström, 1972b; Gupta et al., 1973; Joshi and Murray, 1974; Oya et al., 1974; Small and Watkins, 1975).

Although glycosidases and exopeptidases are incapable of dissolving rabbit blastocyst coverings in vitro and presumably only support the lysis, *endopeptidases* are entirely capable of a complete desintegration of the coverings (see 3.2.3.; Tab. 10). The coverings of rodents (here zona pellucida), which have been especially well researched in this respect, are also easily dissolved by various proteinases (Smithberg, 1953; Chang and Hunt, 1956; Conrad et al., 1971; Mintz, 1971; Mintz and Gearhart, 1973). Papain and trypsin have proven especially effective against rabbit blastocyst coverings (see 3.2.3.). On the other hand, a remarkable resistence of the zona pellucida to papain is observed in the mouse, rat and hamster as well as in the unfertilized and fertilized eggs of rabbits (Smithberg, 1953; Chang and Hunt, 1956). This is doubtless a consequence of the differing chemical composition of blastocyst coverings and zona pellucida. It is interesting that rabbit blastocyst coverings are well lysed by elastase but not by collagenase and plasmin (in concentrations up to 0.5—1.0 mg/ml). Collagenase is also ineffective against the zona pellucida of mouse blastocysts (Bowman and McLaren, 1970).

In considering the especially strong effect of papain it must be kept in mind that the cysteine added to activate the enzyme could have increased the effect by cleavage of S-S bridges. In rodents it is possible to dissolve the zona pellucida in vitro by SH compounds alone (Inoue and Wolf, 1974a and b), and, in contrast to the rabbit, by a fairly acid milieu as well (Brun and Psychoyos, 1972; Inoue and Wolf, 1974a and b) which emphasizes that the rodent coverings are in general less resistant than those of the rabbit.

As endopeptidases not only lyse the blastocyst coverings in vitro, but can also be found in quantity in the endometrium, uterine secretion, blastocyst coverings and blastocyst tissue (see 3.2.2.2.2.), we must discuss them in more detail.

4.4.2. The Role of Endopeptidases (Proteinases) in the Dissolution of the Blastocyst Coverings and in Implantation

4.4.2.1. Biochemical Characterization of Blastocyst and Uterine Secretion Proteinases in the Rabbit

The biochemical characterization of these enzymes is still in its infancy; this is principally due to the difficulty in getting adequate amounts of enzyme material from mammalian embryos. Demonstration of blastocyst proteinases (blastolemmase, entoderm proteinase, see 3.2.2.2.2.) in the rabbit has only met with success using gelatin films as a substrate (Denker, 1969, 1971d, 1972, 1974a, 1975, 1976a and b; Kirchner, 1972a). Other proteinase substrates (chromogenic trypsin and chymotrypsin substrates, elastase substrates) were, in as far as we tested, not cleaved by blastolemmase at any significant rate (see 3.2.2.2.2.). Only small amounts of TCA-soluble peptides were released from casein. This speaks for a considerable specificity in that the enzyme only cleaves a few bonds in the protein molecule. The catalytic specificity will be studied in more detail as soon as pure samples of the enzyme are available; purification work (using affinity chromatography) is in preparation.

The most important statements made to date about the biochemical classification of blastolemmase have been derived from *experiments with specific inhibitors,* especially with active site-directed reagents (see 3.2.2.2.2.). The inhibition by inhibitors with a high specificity for *trypsin* (some of them react exclusively with trypsin from certain species) such as PSTI, SSPI, NPGB and chicken ovomucoid, show clearly that the active site of the enzyme must have great similarity to that of trypsin. It is similar in this respect to the sperm proteinase *acrosin* which is essential in the penetration of the zona (Fritz et al., 1974a, 1975b). Interestingly, neither enzyme is inhibited by DSI which strongly inhibits trypsin and chymotrypsin. In other respects, though, there are clear differences between acrosin and blastolemmase: blastolemmase does not measurably cleave acrosin substrates like BANA, which points to a more limited substrate specificity for blastolemmase.

Considering the total spectrum of effective inhibitors for blastolemmase we must obviously classify it in the *trypsin family* (Tab. 5) and therefore it is to be looked at as a serine protease (serine at the active site) even though an inhibition by the classical reagent for serine enzymes, DFP, is only weakly demonstrable in the substrate film test most probably due to methodological difficulties (volatility of DFP, see 3.2.2.2.2.). For these reasons we can exclude a classification to other groups of proteinases such as the metalloenzymes (collagenase). The chelating agent, EDTA, is without effect here which further supports this conclusion.

Within the trypsin family blastolemmase can be *differentiated from trypsin and acrosin* by the substrate specifity already mentioned. It is also different from *elastase* or the elastase-like enzymes from human leucocytes as it does not measurably cleave synthetic or natural elastase substrates (see 3.2.2.2.2.) and is not inhibited by DSI which inhibits elastase so strongly (Fritz and Hochstrasser, 1976). Although the spectrum of effective inhibitors largely resembles that of trypsin more than that of chymotrypsin, histochemical substrate film tests on cryostat sections show, interestingly, a weaker but clearly existant inhibition by specific *chymotrypsin* inhibitors (α_1-antichymotrypsin, chymostatin). It could be that this is caused by a chymotrypsin-like enzyme of the trophoblast which may have an auxiliary function in the histochemical test. After the in vivo application of antipain we see that the strong activity of the trophoblastic knobs here is not completely inhibited by blastolemmase inhibitors (see 3.3.2.1. and 4.4.2.2.2.). This could be a result of a compensatorily enhanced production not only of blastolemmase but of this additional enzyme as well. The identity of the latter is not certain, however, and it cannot be excluded that a cathepsin-B-like enzyme (see below) is included. The fascinating question of whether or not we are dealing with a blastolemmase activating enzyme can only be answered by enlarged biochemical investigations.

No clear *pH optimum* can be recognized for blastolemmase in the histochemical gelatin substrate film test. This is certainly partially a consequence of the fact that only an imperfect control of the pH is possible in this test. It must be kept in mind here, however, that several enzymes could be involved in the reaction. Although the gelatinolysis in the abembryonic blastocyst region is especially strong at pH 8.0−8.5, we still find activity in wide regions at the surface of the trophoblast at acid pH values down to 6.0 (Tab. 9).

In *agar gel electrophoresis* performed at pH 8.2, the main gelatinolytic enzyme of the blastocyst tissues, like that of the blastocyst coverings, migrates to the anode. The main activity of the uterine secretion from the same stage (6 2/3 d p.c.) migrates, however, to the cathode (see 3.2.2.2.2., Fig. 17). This corresponds largely to the findings of Kirchner (1972a), although he did not investigate blastocyst tissue and blastocyst coverings separately, but used scrapings from the trophoblast-covered antimesometrial uterine wall 7 1/2 d p.c. In some cases a fraction could be demonstrated in the uterine flushings which, like blastocyst proteinase (blastolemmase), migrated to the anode. Since these uterine flushings came from uteri which contained blastocysts, it is likely that this fraction was blastocyst-derived blastolemmase. That would mean that the *blastolemmase does not originate in the uterine secretion*.

Blastolemmase activity can be more precisely located and can be differentiated from BANA- (trypsin substrate) or GPNA- (chymotrypsin substrate) cleaving fractions using *PAA disc electrophoresis* in microscale (see 3.2.2.2.2., Fig. 16). *Blastolemmase* migrates between β-glycoprotein and albumin, but nearer the albumin fraction. A separation from uterine secretion proteinase is not as successful with this method as with agar gel electrophoresis: the major activity of the uterine flushings also migrates between β-glycoprotein and albumin, but apparently not so close to albumin. The uterine secretion proteinase and the trophoblast proteinase can be clearly *separated from β-glycoprotein* using micro disc electrophoresis; a comparison to the results obtained with agar gel electrophoresis is somewhat problematical as the micro disc gels only contained those fractions which migrate to the anode. Blastolemmase is unstable at acid pH (see 3.2.2.2.2.), thus we are unable to test migration under this condition. The observation that the gelatinolytic fraction of the uterine secretion cleaves BANA, while that of the blastocyst tissue and blastocyst coverings does practically not, is further proof for the argument that we are dealing with two *different enzymes*. These results stand in direct opposition to statements made by Kirchner et al. (1971) wherein β-glycoprotein is designated as the gelatinolytic proteinase of the uterine secretion and as

Table 16. Synopsis of the biochemical properties of the main endopeptidases found in the uterus and blastocyst of the rabbit and the cat

Enzyme	Major site	pH optimum	Inhibitors	Activation by cysteine	Substrate specificity and further biochemical properties
Rabbit					
1. blastolemmase	trophoblast, blastocyst coverings	alkaline	trypsin inhibitors including highly specific ones (like PSTI)	ϕ	serine enzyme, active site trypsin-like, not metal or SH-dependent. Narrow substrate specificity, no measurable hydrolysis of low molecular weight ester or amide substrates
2. trypsin-like uterine secretion proteinase	uterine secretion	alkaline	like 1.	ϕ	largely similar to 1., but hydrolysis of low molecular weight amide substrates (BANA); electrophoretically separable from 1.
3. cathepsin-B-like proteinase	entoderm, endometrium (stroma cells, deep parts of crypts)	acid	iodoacetamide chymostatin antipain	+	SH-enzyme, not metal-dependent. No cleavage of BANA or GPNA detected
Cat					
4. cathepsin-B-like proteinase	trophoblast, endometrial glands and uterine secretion (implantation site)	acid	iodoacetamide chymostatin (antipain)	+	like 3.
5. trypsin-like stroma cell proteinase	endometrial stroma cells	alkaline	trypsin inhibitors like 1.	(+)	trypsin-like serine enzyme, not metal or SH-dependent (?)

the implantation proteinase of the rabbit. An equivalation of glycoprotein with a proteinase seems unlikely if only because of the considerable quantity of it which occurs in the uterine secretion of the rabbit. Kirchner et al. used macro disc electrophoresis which, at 6–7.5 % PAA, has less sieve effect and therefore other separation properties than the micro disc electrophoresis (20 % PAA). In order to localize the proteinase activity these authors laid the unsectioned gel on gelatin films. The inclusion of gelatin in the gels (see 2.4., 3.2.2.2.2.), which allowed a better localization of the proteinase activity, enabled us, however, to separate the gelatinolytic uterine secretion proteinase activity from β-glycoprotein even using macro disc electrophoresis.

A biochemically totally different proteinase activity can be found in the *entoderm*, in the *stroma cells* and in the deeper regions of the *endometrial crypts*, which is so strong that (at least in some stages, see 3.2.2.2.2.) it can be easily detected with the gelatin substrate film test. Investigations made towards an understanding of its biochemical properties have yielded clear evidence that we are dealing with a *cathepsin* from the *B-group* (EC 3.4.22.1). In favor of this assumption are: a pH optimum in an acid region (Tab. 9); inhibition by chymostatin, antipain and iodoacetamide; activation by cysteine, no inhibition by typical trypsin inhibitors (Tab. 7) (see Snellman, 1969; Perlmann and Lorand, 1970; Umezawa, 1972; Barrett, 1974). No greater similarity to trypsin could be determined: BANA was not cleaved in measurable quantities (nor was LeuNA or GPNA). This enzyme is certainly different from cathepsin D which is found in the uterus of many species (Wood et al., 1969; Wood and Barley, 1970; Sapolsky and Woessner, 1972; Woessner, 1973; Moulton, 1974), because of a lack of inhibition by pepstatin and because of the strong inhibition caused by iodoacetamide.

We *conclude* (see Tab. 16) that the main proteinase fraction of the implanting rabbit blastocysts can be separated from the proteinases of the uterine secretion. Because of the apparent function of this enzyme in the lysis of the blastocyst coverings (see 3.2.2.2.2., 3.3. and 4.4.2.2.2.) we call it *blastolemmase* (from "blastolemmata" = general term for mucolemma, oolemma and gloiolemma, see 3.1. and 4.2.). Blastolemmase is, therefore, an enzyme which is in many respects similar to trypsin, but which can be differentiated from it as well as from other members of the trypsin family of enzymes, like acrosin, elastase, elastase-like enzymes of the leucocytes and chymotrypsin on the basis of substrate specificity and its spectrum of effective inhibitors. Whether or not the enzyme is synthesized by the blastocyst (trophoblast) is not certain (see Denker, 1974b, 1975; Denker and Hafez, 1975). Another possibility would be that it is secreted by the endometrium as a proenzyme and is taken up by the blastocyst and activated. In the *uterine secretion* we find an enzyme very similar to trypsin which shows activity with BANA. The trophoblast of the rabbit has, in addition to blastolemmase, a smaller activity of a *cathepsin*-like (and/or a chymotrypsin-like) enzyme. The high proteinase activity of the *entoderm* and the *stroma cells* (as well as the weaker activity of deeper regions of the endometrial crypts) is caused in the rabbit by a cathepsin-B-like enzyme.

4.4.2.2. Inhibition of the Dissolution of the Blastocyst Coverings and of Implantation by the in vivo Application of Proteinase Inhibitors

4.4.2.2.1. Chemistry and Biochemistry of the Inhibitors Used

Four inhibitors were chosen for the in-vivo application which inhibit the blastolemmase of the rabbit well in vitro (Trasylol®, antipain, NPGB, SSPI). ε-Aminocaproic acid (EACA) was chosen as a control as it does not inhibit blastolemmase in vitro (see 3.2.2.2.2.).

Trasylol[®] is a proteinase inhibitor which occurs in considerable quantity in several bovine organs and can be isolated industrially. It is a basic polypeptide with a molecular weight of ca. 6500, whose structure is known (Trautschold et al., 1966; Vogel et al., 1966; Werle, 1969, 1972). Trasylol[®] is barely toxic even at high doses, the LD_{50} for the rabbit with i.v. injection is 500 000 KIU/kg, which is about 90 mg/kg for the preparation we used. This was one of the primary reasons we chose this inhibitor for our experiments. The inhibition spectrum is relatively wide: it inhibits trypsin, chymotrypsin, various kallikreins, plasmin etc., but acrosin only weakly (Fritz et al., 1975b). Blastolemmase is strongly inhibited (see 3.2.2.2.2., Tab. 5). At pH 7.8, the inhibitor constant for trypsin is $K_i = 2 \cdot 10^{-11}$ Mol/l (Vogel et al., 1966).

Antipain has a narrower spectrum compared to Trasylol[®]: it only inhibits trypsin-like enzymes, but not chymotrypsin. Blastolemmase is effectively inhibited (see 3.2.2.2.2., Tab. 5). It is, like Trasylol[®], a natural inhibitor, and is harvested from filtrates of Actinomycetales (Umezawa, 1972; Wingender et al., 1975). Antipain is a low molecular peptide of 4 amino acids with a urea group and, terminally, arginal. It has a molecular weight of ca. 605, only about 1/10 of that of Trasylol[®] Antipain, like other related bacterial inhibitors, has recently found interest in proteinase biochemistry as a tool for the study of the interaction between proteinases and inhibitors. Because of its low toxicity (no death in the mouse at i. v. injections of 125 mg/kg; Ld_{100} = 250 mg/kg), possible therapeutic uses are being considered (Umezawa, 1972).

Boar seminal plasma inhibitor (SSPI) (Fritz et al., 1975a, 1976a) is also a natural inhibitor with polypeptide character. The preparations used in our in vivo experiment contained fractions I and II (MW 6000–13.000). It strongly inhibits blastolemmase (see 3.2.2.2.2., Tab. 5). Its spectrum of effectivity is similar in many respects to that of antipain, and it is largely identical to that of an inhibitor from human seminal plasma (HUSI-II) (Fritz et al., 1975a).

p-Nitrophenyl-p'-guanidinobenzoate (NPGB) has found use in biochemistry as an active site titration reagent for trypsin-like enzymes (Chase and Shaw, 1970). As the dissociation of the acyl enzyme complex formed takes place slowly, NPGB is also an effective inhibitor. It is especially interesting on account of its low molecular weight which enables it to pass cell membranes, as in the case of acrosin in living sperm, for example, where the acrosomal membranes protect the enzyme from attack by the majority of inhibitors. NPGB, however, is able to inhibit acrosin even in intact, living sperm (Wendt et al., 1975b; Fritz et al., 1976a). In many laboratories, NPGB and similar inhibitors are recently under investigation for a possible contraceptive use (unpublished). One must consider, however, that the p-nitrophenol which is released can have side effects. Blastolemmase is very effectively inhibited by NPGB (see 3.2.2.2.2., Tab. 5).

ε-Amino caproic acid (EACA) does not inhibit blastolemmase (see 3.2.2.2.2., Tab. 6). This inhibitor is used therapeutically as an inhibitor of plasminogen activators similar to the more effective AMCHA (Andersson et al., 1965; Bang, 1971; Witt, 1975). Direct inhibition of plasmin takes place to a lesser extent. The inhibitor was used as a specificity control for the in-vivo experiments.

4.4.2.2.2. Demonstration of the Key Role of Proteinases in Implantation Using in vivo Inhibitor Experiments

The in vivo inhibitor experiments we carried out showed clearly that *proteinase inhibitors which inhibit blastolemmase in vitro, handicap or prevent the in vivo dissolution of the blastocyst coverings and implantation* (see 3.3.2.1., Tab. 12). It is well known of the inhibitors chosen here that they possess a high specificity for certain proteinases (see above). In our experiments with the dosage used and with intrauterine application, we saw *no signs of toxic side effects on the mother.* The implantation and development of the embryos in the control uterus of the same animal progressed undisturbed. The injection of an inhibitor which does not inhibit blastolemmase, i. e. EACA, had no influence on implantation, even at molar dosages 500 times higher than that of Trasylol[®] (6 mg ~5 · 10^{-2} mMol EACA compared to 0.6 mg ~10^{-4} mMol Trasylol[®], see 3.3.2., Tab. 11 and 12). The slight disturbance of the lysis of the coverings and attachment which was observed after the extremely high dose of 12 mg (about 10^{-1} mMol) EACA/uterus might be attributed to unspecific effects.

We conclude that the observed inhibition is actually caused by an inhibition of blastolemmase or similar proteinases, and that proteinase activity plays an important role in the initiation of implantation.

It is characteristic for *Trasylol*® and *antipain* treatment that the blastocysts are not so strongly damaged that they immediately die. They also do not collapse which they often do after various other types of damage. They actually continue to *expand* although at a reduced rate. The *lysis of the coverings is inhibited,* however, and implantation does not get started. The inhibition of the coverings lysis is more pronounced in the case of Trasylol® than with antipain. After the application of a single dose of Trasylol® a regular lysis appears to no longer be possible up to 8 1/2 or 9 1/2 d p.c. Only under the electron microscope is it possible to detect a slight erosion of the surface. Finally the coverings, which have been stretched further and further during expansion, tear apart (as is also observed in in vitro cultures) and their remains are pushed to the side by the constant movements of the endometrium against the blastocyst (Böving, 1952, 1959, 1960).

Practically no *blastolemmase activity* can be detected after treatment with Trasylol®. After treatment with antipain, however, low activity can be detected 7 1/2 d p.c. on the surface of the trophoblast and in the abembryonic coverings. The lamination of the latter disappears, and finally local lysis begins over some of the *trophoblastic knobs* (mostly located somewhat laterally). After antipain treatment (and hints of it also after the application of 0.6 mg Trasylol®) a clear and often even a strong proteinase reaction can be detected in the trophoblastic knobs (Fig. 27b, 30b), which is not the case in untreated controls (see 3.3.2.1. and 3.2.2.2.2). This activity is most obvious at 8 1/2 d p.c. and is *maximal at the center of the abnormally enlarged trophoblastic knobs which have been prevented from attachment.* The suspicion arises that the application of inhibitor has triggered a *compensatorily enhanced production* of proteinase by the trophoblastic knobs. Interestingly, the proteinase activity of these trophoblastic knobs exhibits certain biochemical differences compared to blastolemmase of untreated blastocysts: it is only incompletely inhibited by SBTI, NPGB and ovomucoid. There are two possible explanations for this: either the enzyme is partially protected from attack by the inhibitors in the enlarged trophoblastic knobs by compartmentalization or by storage as a proenzyme, or that we deal in fact with a different proteinase which may be formed also in normal pregnancy, but which in that case recedes quantitatively in the background during the substrate film test. It may nonetheless work synergistically or as an activator. It should be kept in mind that with micro disc electrophoresis several BANA or GPNA cleaving fractions are demonstrable even in untreated blastocysts and outside of the migration region of blastolemmase (see 3.2.2.2.2., Denker and Petzoldt, 1977).

After treatment with Trasylol®, as with antipain, beginning 8 1/2 d p.c. we observe the delayed attachment of single, abnormally few trophoblastic knobs to the endometrium. Thereafter some embryos successfully initiate mesometrial implantation (see Tab. 12). We have here the interesting situation before us of a *delay of implantation* which does not otherwise occur in the rabbit. Other experimental procedures, for example influencing hormonal control (ovariectomy, see Denker, 1972; delayed secretion, see Beier, 1974a and b; Beier and Kühnel, 1973), which can also hinder implantation, are usually followed by rapid degeneration of the blastocysts. Apparently treatment with proteinase inhibitors is more *specific,* as, apart form the process of implantation, other physiological processes are not seriously affected even though Trasylol®,

antipain, SSPI and NPGB are quite capable of inhibiting a number of various tryp-
sin- (or chymotrypsin-) -like enzymes (literature, see Tab. 5). As the example of the
cathepsin-B-like entoderm proteinase shows, the activity of some of the proteinases
different from blastolemmase, which are perhaps important in the metabolism of the
blastocyst, is unaffected in vitro by Trasylol® and SSPI (see 3.2.2.2.2.). Although antipain inhibits in vitro the cathepsin activity of the entoderm, it apparently does not
reach this tissue in significant quantities as the enzyme is demonstrable in the blastocysts of the treated animals. A blockade of the system's entire arsenal of catheptic enzymes cannot be assumed anyway. The inhibitors applied do not, for example, inhibit
cathepsin D which, although it did not appear in our gelatin film test, is present in the
uterus of many species (literature see 4.4.2.1.).

The prerequisite for an eventually successful, though delayed and locally restricted attempt at attachment by some embryos is possibly fulfilled in that the *intrauterine concentration of inhibitors*
has in the meantime considerably decreased (see 3.3.2.3.). The number of determinations which
have been carried out are certainly not adequate to allow us to formulate the kinetics of elimination. This must be the object of a more complete investigation. For this reason we have restricted
ourselves to giving single values and averages. It is apparent that the values deviate widely, even
when the uterus was flushed immediately after application. Apparently varying amounts of inhibitors are immediately adsorbed in the uterus or neutralized. The values obtained after immediate
flushing deviate even more widely when they are calculated on the basis of protein content and not
on the basis of volume units of flushings (see Tab. 15). We calculated on the basis of protein content because in the postimplantation stage it is not possible to flush the entire uterus as it was at
the beginning; it is now possible only in those regions lying between the blastocysts. Unfortunately
the protein content of the uterine cavity changes not only from phase to phase, but varies tremen-
dously in individuals as well (Beier, 1968a; Kulangara, 1972), which explains the wide deviation
found. Calculation on the basis of DNA is, on the other hand, not possible with secretions. Despite
these limitations, the results show clearly (Fig. 33, 34) that *within the first 24 hours the activity
of inhibitor falls steeply* and, as expected, approaches zero asymptotically. In the case of Trasylol®,
as opposed to antipain, clear remnants of activity can be detected 24 and 48 hr after application.

The relatively long *lingering* of *Trasylol®* activity in the uterine lumen is noteworthy. Presumably the fact that Trasylol® is strongly basic plays a role here and it may well be bound to the acid
mucosubstances found abundantly in uterine secretion and at the surface of uterine epithelial cells
(Denker, 1970a and b). It is particularly probable that this *inhibitor collects in the blastocyst coverings* which are largely made up of acid mucosubstances. This could be significant in that the blastocyst coverings lysis is so effectively inhibited by small doses of Trasylol®. We would offer yet
another explanation for the lingering of Trasylol® in the uterus. Possibly Trasylol® is bound similarly to the way it is bound in the kidneys and stabilized in this high molecular form (Werle, 1969);
the necessary cleavage of terminal amino acids could be carried out by exopeptidases in the uterine
secretion, for example cleavage of N-terminal arginine by the high arylamidase activity in uterine
secretion (see 3.2.2.2.1., 4.4.1.).

The result of the delayed attempt at attachment cannot be compared to normal implantation, possibly because a remaining activity of inhibitors hinders or perhaps because in the meantime the *morphological condition of the uterine epithelium has
changed* and has become unfavorable for normal attachment. The investigation of
semi- and ultrathin sections shows impressively that the transformation of the antimesometrial cavum epithelium into continuous wide symplasms proceeds similarly to
the control side, despite no attachment of the blastocyst at first. We do see, however,
that the trophoblastic knobs can also fuse with these symplasms as soon as the coverings have disappeared (through rupture in the case of Trasylol® or partial lysis and
rupture in the case of antipain) (see 3.3.2.1.). Whether or not all trophoblastic knobs
are capable of this fusion or whether only certain ones have this capability as Steer
(1970b, 1971a and b) postulates, is difficult to say. Although in the normal situation
the relatively large trophoblastic knobs each fuse with only a few uterine epithelial

cells at first (see Böving, 1963, 1970; Enders and Schlafke, 1971; Schlafke and Enders, 1975), they are confronted with a broader symplasmatic mass after the delay caused by treatment with inhibitors. One has the impression that the *trophoblast is disadvantaged quantitatively:* the few nuclei and the small amount of trophoblastic knob cytoplasm dive into an ocean of uterine epithelial symplasm (Fig. 23). It is conceivable that in this situation the *establishment of contact with the maternal blood vessels cannot take place* in the normal fashion: according to Böving, the trophoblastic knobs always attach to those uterine epithelial cells which lie directly over a maternal subepithelial capillary (see 3.1.1., Fig. 4). Böving (1954, 1963) showed that such cells are distinguished from other cells metabolically, and he postulated that they canalize the attachment of the trophoblastic knob at exactly this spot by a localized metabolite transport (whose nature as CO_2-bicarbonate transport remains questionable, see Böving, 1963, 1970) and establishment of the consequent gradients. This allows the trophoblastic knob to find the shortest possible route of contact to the maternal circulation. The clear parcelling of the metabolite transport and therefore *the directing of trophoblastic knob growth* in the direction of subepithelial capillaries can no longer function as soon as the lateral cell walls of the uterine epithelial cells have ceased to exist.

If this concept holds true, one may expect that after treatment with inhibitor, because of the delay in attachment, the trophoblastic knobs occasionally attach to the endometrium between subepithelial capillaries, which would *call into question the functional efficiency of this type of contact for effective transport* between the maternal blood stream and the embryo. As a matter of fact, the *transport of protein into the blastocyst* is clearly disturbed after treatment with proteinase inhibitors (see 3.3.2.1.). Normally the blastocyst fluid contains little protein until 7 d p.c. (Zimmerman et al., 1963; Zimmermann, 1965; Hafez and Sugawara, 1968, Hamana and Hafez, 1970; Lutwak-Mann, 1971; Petzoldt, 1974). After 7 1/2 d p.c. one finds traces, at 8 d p.c. large amounts, which can be seen easily even in tissue sections. These are largely plasma proteins (including fibrinogen) which pour into the blastocyst from the maternal blood stream during this "open phase" (Brambell and Hemmings, 1949; Zimmerman et al., 1963; Zimmermann, 1965; Kirchner, 1969). In those blastocysts treated with Trasylol[®] and antipain the blastocyst cavity remains free of protein, at least as far as could be seen in tissue sections, until 11 1/2 d p.c. (after antipain treatment traces of protein can be often found 9 1/2 d p.c.). This emphasizes our comment that the incomplete contact established after 8 1/2 d p.c. is functionally inferior.

Perhaps it is because of the resulting "nutritional deficiency" that the *embryonic anlage (embryonic disc)* of blastocysts treated with inhibitor degenerates (see 3.3.2.1., Tab. 13). In contrast to the embryonic anlage the *extraembryonic tissues* develop apparently well at first despite treatment with inhibitor. The trophoblast and entoderm cells are somewhat flatter than in the control blastocysts, but all of their typical organelles and inclusions (even the protein crystals of the trophoblast) are present. The number of large, protein-containing granules per area even appears raised in many regions of the trophoblast. The (extraembryonic) entoderm develops its specific proteinase activity which even appears increased (compensatorily?) (Fig. 22).

The effects of treatment with *NPGB* have to be discussed separately as not only is the lysis of the blastocyst coverings very effectively inhibited, but in some of the embryos an early and widespread *damage to the blastocyst tissue* is observed (see 3.3.2.1.). As early as 7 1/2 d p.c. we find a disorderly clump of largely degenerated cells in the collapsed blastocyst coverings of these embryos which show no signs whatsoever of lysis or blastolemmase activity (Fig. 31). Interestingly some of these cells still show proteinase activity which is probably, because of the spectrum of effective inhibitors, catheptic entoderm proteinase. The entoderm cells therefore appear to have survived the treatment with NPGB fairly well.

As the before-mentioned proteinase inhibitors, despite a clear inhibition of implantation, do not lead to collapse of the blastocysts, but rather leave the tissue relatively unharmed so that expansion can even take place, we must assume that NPGB has a toxic effect on the blastocyst possibly because of the release of p-nitrophenol (see 4.4.2.2.1., Tab. 12). The toxic damage is lethal in a certain percentage of the cases. The rest of the blastocysts which overcome this damage, however, develop a normal blastolemmase activity as early as 7 1/2 d p.c. and implant normally. One must keep in mind here that for this reason the danger exists that NPGB or similar substances could produce *birth defects*. This possibility always exists where embryos are only slightly damaged but do not die (so-called "Durchbrenner", Gottschewski and Zimmermann, 1973). One must also consider a toxic effect on the mother although no such effect was observed in the experiments done.

The *boar seminal plasma inhibitor (SSPI)* at in vivo doses of 6 mg per uterus showed only moderate inhibition of the blastolemmase activity and coverings lysis (see 3.3.2.1.). Higher doses were impossible to test as this inhibitor cannot easily be isolated in such quantities. The inhibitor properties of SSPI are largely identical to those of HUS1-II, a human seminal plasma inhibitor (Fritz et al., 1975a, 1976b; Schiessler et al., 1976). It can therefore be assumed that the small quantities of HUS1-II which occur in human ejaculate do not interfer with the implantation of the blastocyst inasmuch as this process in man possibly depends on an enzyme similar to the blastolemmase of the rabbit. HUS1-I, another inhibitor from human seminal plasma, has an inhibition spectrum similar to DSI and therefore, like DSI, should not inhibit blastolemmase.

Summarizing the results of the in vivo inhibition experiments we arrive at the conclusion that *trypsin-like proteinase activity* plays a decisive role in the *blastocyst coverings lysis* of the rabbit. The same can be said somewhat less definitely for the *attachment of the trophoblast* to the endometrium, as this is to a certain extent still possible, though delayed, after treatment with inhibitor, as soon as the coverings are removed by lysis (antipain) or rupture (Trasylol®). This raises the question whether the proteinases are directly involved in the initiation of attachment or only play an indirect role. This likewise holds true for the uptake of protein into the blastocyst cavity which is markedly disturbed after treatment with proteinase inhibitors (see above). The total picture we get from analysis of normal pregnancy (see 3.1. and 3.2., 4.1. and 4.2.), the models derived therefrom (see 4.3.), from in vitro experiments on the enzymatic dissolution of the blastocyst coverings (see 3.2.3., 4.4.1.) and from in vivo experiments with blastocyst models (see 3.3.1.1., 4.6.) and hormonal regulation (see 4.5., Denker, 1972), leads us to believe that the proteinases play a significant part in the dissolution of the blastocyst coverings. The ability of the blastocysts to survive and even continue expansion is evidence for the supposition that it is primarily this process which is disturbed by the in vivo treatment with proteinase inhibitors. As uterine secretion proteinases as well as the blastolemmase of the blastocysts are inhibited by these inhibitors, a final judgement on the *role of the endometrium or blastocyst* during implantation cannot be made on the basis of these experiments. Hopefully, inhibitors will be found which are able to selectively inhibit blastolemmase or uterine secretion proteinase. We have prepared experiments to test inhibition by specific antibodies.

4.4.2.3. Endopeptidases Related to Implantation in Man and in Other Mammalian Species

The proteinase activity which can be found at implantation sites in the cat (see 3.2.2.2.2.) shows remarkable similarities in its histochemical distribution pattern and its phase-specific occurrence to the blastolemmase activity of the rabbit. In this species one does not, however, observe the main activity in the dissolving blastocyst coverings (here: zona pellucida) but in the trophoblast and on the uterine epithelial surface. Even if it has been less intensively investigated than in the rabbit (especially because of the great difficulty in obtaining enough material from cats in early pregnancy) a study of the biochemical properties shows that the main gelatinolytic enzyme of the trophoblast is not related to the blastolemmase of the rabbit. We are dealing instead with a SH-group dependent *cathepsin* (B group) which is very similar to the cathepsin of rabbit entoderm and stroma cells as well as that of deeper regions of the endometrial crypts (see 4.4.2.1., Tab. 16). It is at present impossible to decide whether the blastocysts of the cat in addition to showing this strong cathepsin activity also contain *blastolemmase* in lesser quantities. 12 d p.c. when the zona is being dissolved and the cathepsin activity is not yet as high as it will be 14 d p.c., the pH optimum of trophoblast proteinase as found with the histochemical gelatin film test is closer to neutrality compared to the values for 14 d p.c. (Tab. 9). This could be an indication for the presence of a blastolemmase-like enzyme. Perhaps the cathepsin activity predominates so heavily that a demonstration of trypsin-like enzymes is made difficult.

In contrast to blastolemmase, the site of action for the cathepsin-B-like enzyme is, because of the pH optimum, more likely to be found *intracellularly* (lysosomes). It seems unlikely, therefore, that this enzyme is important for the lysis of the blastocyst coverings inside the uterine lumen where the pH is presumably alkaline as in the rabbit. It is remarkable, however, that such a high gelatinolytic cathepsin activity is found, which dissolves gelatin membranes even faster than the peak blastolemmase activity of the rabbit, and that this activity is especially high during the trophoblast's invasion of the endometrium. We should mention here that cathepsin B is found in large amounts in *tumor cells* and that a certain correlation exists between the presence of this enzyme and the tendency towards invasive growth (Sylvén et al., 1974). With regard to implantation, on the basis of the little data we have at present, no correlation could be seen, however, between the type and intensity of the proteinase activity and the histological *type of placentation,* as has often been assumed (for discussion see Gräfenberg, 1910; Mossman, 1937; Billington, 1971; Starck, 1975; Steven, 1976): in the cat we see a higher activity of endopeptidases demonstrable in the gelatin film test than in the rabbit, although here the maternal tissue is less deeply invaded than in the latter (endotheliochorial as opposed to hemochorial placenta).

Comparable proteinase investigations have only been carried out in a few other species. The enzyme of the *uterine secretion* appears to play a special role in the *mouse* and *rat.* An enzymatic activity found in the uterine secretion of these species in dependence on the "estrogen surge" shortly before implantation leads to lysis of the zona pellucida not only in living blastocysts but in dead embryos as well. It has been postulated that this factor is identical with a casein-hydrolyzing proteinase activity which occurs simultaneously in the uterine secretion and that this proteinase is essential for the initiation of implantation (*"Implantation-Initiating Factor";* Mintz, 1971, 1972; see also McLaren, 1970; Pinsker et al., 1974). Considerable confusion was caused by the observation that mouse blastocysts in vitro and in delayed implantation (through lack of the estrogen surge) are capable of shedding their zona pellucida by mechanical rupturing (hatching) (McLaren, 1967, 1970; Cole, 1967; Bitton-Casimiri et al., 1970; Rumery and Blandau, 1971; Bergström, 1972a). In vivo, during normal pregnancy, the zona pellucida seems to be lysed, however, also in this species (Potts and Wilson, 1967; Reinius, 1967; Smith and Wilson, 1974).

On the basis of the cited investigations we are unable to entirely exclude the possibility that the trophoblast of the mouse blastocysts also produces enzymes which, at least in vivo, take part in the lysis of the coverings. *Trypsin- and chymotrypsin-like enzymes* have been demonstrated not only in the endometrium and uterine secretion, but in mouse blastocysts as well (Andary et al., 1972;

Andary, 1974; Andary and Dabich, 1974). Interestingly these authors found in PAA electrophoresis in addition to BANA-cleaving fractions another trypsin-like fraction (inhibited by TLCK) which does not hydrolyze this synthetic substrate but does cleave protein. This fraction, which is perhaps best compared to the *blastolemmase* of the rabbit, has not been more closely investigated because the above authors largely carried out their experiments with BANA or GPNA which allow kinetic assays.

Interesting in *vivo inhibition experiments* have been carried out in the mouse by the same authors. These are the only previously existant experiments which are similar to our in vivo experiments. They used, however, mainly the inhibitors TLCK and TPCK including only few experiments with the basic trypsin inhibitor from bovine organs (Kunitz, i. e. Trasylol$^{®}$). TLCK and TPCK are problematical for several reasons: they possess unfavorable kinetic properties on the one hand (relatively slow formation of the inhibitor complex, therefore only slight inhibition by TLCK in the substrate film est, see 3.2.2.2.2., Tab. 5); they are not water-soluble and must therefore be dissolved in organic solvents, and they are alkylating agents making it difficult to exclude possible side-effects in vivo. All three inhibitors were either injected into the uterine lumen or given in "slow release devices". All three inhibitors, using either method of application, led to a drastic decrease of the number of implanting and developing embryos (Dabich and Andary, 1974). No further analysis was carried out, however, to see whether it is in fact the first phases of implantation which were influenced or whether it is the further development which was disturbed; only the number of apparently normal fetuses was determined between 12 and 18 days.

Bergström (1970), using a histochemical substrate film test based on gelatin, was unable to demonstrate any proteinase activity of mouse blastocysts undergoing implantation; only the uterine epithelium showed some slight activity. We were able to confirm this for the same species and for the *Syrian hamster* as well from our own investigations, although the number of experiments carried out was limited (Denker, unpublished).

Blandau (1949) was able to demonstrate a high gelatinolytic activity in implanting blastocysts of the *guinea pig*. It originates apparently in the trophoblast which grows through the zona pellucida at the abembryonic pole with pseudopod-like extensions, so that the zona comes to look like a sieve before it is entirely dissolved (Spee, 1901; Blandau and Rumery, 1957; Blandau, 1971b; Parr, 1973). Unfortunately the later investigations on a greater scale were only carried out on post-implantation phases of this species (Owers and Blandau, 1968, 1971; Owers, 1970, 1971). Only a few investigations of the biochemical characteristics of this enzyme have been carried out. Accordingly, at this early post-implantation phase in the guinea pig, we are probably dealing with a *cathepsin* with its pH optimum in the slightly acid region; it is not inhibited by EDTA. It is apparently similar to the cathepsin we were able to demonstrate in the implanting blastocyst of the cat and in the entoderm (and endometrium) of the rabbit. The question remains open whether or not the pseudopod-like extensions which the trophoblast develops during the penetration of the zona build and secrete a blastolemmase-like enzyme. The morphological evidence mentioned leads us to suppose this. The very small size of the guinea pig blastocysts presents, however, special difficulties for investigation.

Using the same technique in the *rat*, Blandau and Owers (l.c.) found no gelatinolytic activity in the blastocysts. This causes no surprise considering the close relationship between this animal and the mouse who likewise gives a negative result (see above).

Exhaustive investigations based on considerations similar to the present paper were carried out on *human* material (including endometrium, decidua and placenta) by Schmidt-Matthiesen (1963, 1967, 1968, 1970). He used *fibrin* as a substrate in proteinase investigations (fibrin plate test or thrombelastrogram). Schmidt-Matthiesen concluded that the *endometrium* is the major source of enzymatic activity and that the activity is especially high during the secretion phase. During pregnancy he found only a slight fibrinolytic activity of the decidua, which is not a result of any lack of fibrinolytic enzymes, but rather caused by the high activity of *inhibitors*. These observations accord with results from experiments some of which were made already at the beginning of this century (Polano, 1907; Gräfenberg, 1909, 1910; Halban and Frankl, 1910; Frankl and Aschner, 1911; Caffier, 1929a and b; Abe, 1932). Schmidt-Matthiesen emphasized that intact villi from the 3rd to the 4th month of pregnancy are fibrinolytically inactive as are mature placentas (but not under pathological conditions). He concluded that *embryos probably activate a latent enzyme activity of the endometrium rather than producing the fibrinolysin* themselves. In connection with this it is worth mentioning that several nonspecific stimuli raise the activity of lysosomal enzymes in the endometrium (Abraham et al., 1973). The fibrinolytic activity of the uterine secretion is raised in the presence of IUDs. Certainly *plasminogen activators* play a role here and are under investigation in several current investigations. ε-Aminocaproic acid-containing IUDs are being tested, for example (experiments in the human, macaques and rats, see Koutsky et al., 1969; Shaw et al., 1970, 1973, 1975; Liedholm and Åstedt, 1975; Larsson et al., 1975a and b).

It is difficult to make a comparison of these numerous observations with the results described here for rabbit *blastolemmase*. It is probable that the fibrinolytic activity of the endometrium and uterine secretion is due to a completely different enzyme or enzymatic system. Fibrin is a poor substrate for blastolemmase (see 3.2.2.2.2.). The in vivo application of EACA, an inhibitor of plasminogen activation, has no influence on the blastolemmase activity or on implantation in the rabbit (see 3.3.2.1.). A biochemical characterization of the fibrinolytic enzyme or enzymatic system which was investigated by Schmidt-Matthiesen and Shaw et al. is still lacking. Unfortunately there have been no investigations done on the proteinases of the early attachment stages of human implantation or the implantation of other primates, and it is unknown whether or not in this case an enzyme with properties similar to those of blastolemmase is released by the trophoblast. An answer to this question would be of great interest, especially because in the guinea pig, whose mode of implantation resembles that of the human in many respects, the trophoblast very likely does actually release such an enzyme during attachment (see above).

4.4.2.4. Hatching Enzymes of Lower Animals

The eggs of oviparous animals are surrounded with more or less tough extracellular coverings which must be shed if the larva or young animal is to assume the free-living form. There is a certain analogy here to the processes discussed in the overcoming of mammalian blastocyst coverings in that proteinases apparently play a definitive role. The proteinases are clearly formed and secreted by the embryo (or young animal) since this stage of development is usually reached outside of the maternal body. Since such eggs can easily be obtained in large quantities, the biochemical isolation and characterization of these proteinases is considerably easier than in the case of mammals and has been carried out in several laboratories in the last few years.

In amphibians (Xenopus laevis, various species of Rana) (see Carroll and Hedrick, 1974; Katagiri, 1975; Yoshizaki, 1975; Yoshizaki and Katagiri, 1975) the hatching enzyme is formed by intraepithelial gland cells which are aggregated in the epidermis at the head (parietal region) of the embryo. Before hatching takes place, electron micrographs show a collection of large granules in the apical part of these cells whose content is extruded into the perivitelline space at the initiation of hatching; after extrusion the cells die.

The enzyme of Rana is a thermolabile endopeptidase with a molecular weight of 55000 to 60000 and a pH optimum between 7.4 and 7.8. It is inhibited by DFP and TLCK, but not by SBTI. It remains to be cleared up whether or not it belongs to the trypsin family. As EDTA has an inhibiting effect, metal ions are probably essential. In the case of the enzyme from Xenopus the *number of bonds which this enzyme cleaves appears to be limited* (this is probably valid for blastolemmase too, see 4.4.2.1.); it is for this reason difficult to demonstrate in the usual biochemical tests, and its physiological function is apparently *not the complete dissolution but rather the softening of the coverings* (Carroll and Hedrick, 1974; Katagiri, 1975). When enzyme preparations and embryonic coverings of various species are tested, a certain degree of species specificity is observed (Katagari, 1975).

In fishes (Salmo, Oryzias, Fundulus) (Yamamoto, 1963; Kaighn, 1964; Ogawa and Ohi, 1968; Yamagami, 1970, 1972, 1973, 1975; Hagenmaier, 1972, 1974a and b, Yamamoto and Yamagami, 1975) the hatching proteinase is secreted from single-celled glands which are located in the head region and oral cavity. It is an endopeptidase which is apparently not a serine enzyme (no inhibition by DFP or various trypsin inhibitors) but rather a metallo-enzyme (inhibition by EDTA and other chelators). Synthetic trypsin and chymotrypsin substrates (BAEE, BANA, BTEE, N-acetyl-phenylalanine naphthylester) are not cleaved by the Salmo enzyme (Hagenmeier, 1974a); it has been reported that the Oryzias enzyme shows activity with BAPA as a substrate (Yamagami, 1973). The latter appears to be slightly inhibited, however, by pepstatin (Yamagami, 1975). The pH optimum lies between 7.9–9.0; the molecular weight was found to be between 8000–10000 which is unusually low. Kaighn (1964) reported deviating results for the molecular weight (15000 to 40000) and inhibition by DFP, although there was some evidence that DFP was bound outside the active site in this case.

In the *sea urchin*, the hatching process takes place very early in development, namely in the blastula stage. The isolated hatching enzymes (from Strongylocentrotus purpuratus, Hemicentrotus pulcherrimus, Anthocidaris crassispina) (Yasumasu, 1960, 1961, 1963; Barrett and Angelo, 1969) are also endopeptidases which appear not to have serine at their active site (resistance to DFP). No esterase activity could be demonstrated, the pH optimum was near 8.0 and they were Ca^{++}-dependent (Barrett, 1968). Apparently the enzyme synthesis is coded entirely by maternal genes; the transcription possibly takes place as early as during oogenesis (Barrett and Angelo, 1969). No chemical connection to proteinases which are extruded into the perivitelline space during fertilization appears to exist (Carroll and Epel, 1975).

In *insects* a comparatively late appearing hatching enzyme is well characterized, the so-called *cocoonase.* In silk moths (Bombyx mori, Antheraea polyphemus) it is formed by a specialized group of cells on a specific part of the maxilla, the galea, and is secreted at hatching from the cocoon (Hruska and Law, 1970; Kafatos, 1972). It is an enzyme of the trypsin family: inhibited by DFP, TLCK and SBTI, with a pH optimum of 8.0. At this slightly basic pH the enzyme is remarkably stable. It has esterolytic activity and can be titrated with NPGB.

The cocoon which is lysed by cocoonase is naturally not at all comparable to egg coverings. Of the hatching enzymes which dissolve real embryo coverings, those of the sea urchin and of fish belong to an entirely different category than blastolemmase: they are not serine enzymes and therefore have no connection to the trypsin family but are more than likely metalloenzymes. The hatching enzymes of the amphibians, though, could be categorized like blastolemmase as members of the trypsin family.

4.5. The Physiological Regulation of the Process of Implantation

The phase-dependent appearance of the high blastolemmase activity of the blastocyst at the time of implantation requires a regulatory mechanism which is to date unknown. It is known that implantation, and to a certain extent the preimplantation development, is regulated by maternal *steroid hormones.* The *uterine secretion* changes its composition from one day of preimplantation to the next which can be followed from the appearance of individual proteins (see Beier, 1968a, 1970a, 1971, 1974a and b). This is likewise valid for various enzymes and especially impressive in the amino acid arylamidase (Denker, 1969; see 3.2.2.2.1., 4.4.1.). It has been demonstrated that the regulation of this phase-specific change in the uterine secretion is brought about by the varying levels of maternal estrogens and gestagens. For regular embryonic development it is necessary that the uterine milieu exactly corresponds to the developmental age of the embryo, which means that a *synchronization* must be maintained between the developmental processes in the embryo and the phases of pregnancy (or the pattern of secretion) of the endometrium (Chang, 1950; Beier et al., 1972b; Beier, 1973,

1974a and b, Beier and Kühnel, 1973). This synchronization is guaranteed on the one hand by their common starting point: ovulation and subsequent fertilization is the beginning for the development of the embryo as well as for the development of the corpus luteum and hence progesterone secretion. In addition, the synchronization is secured by interactions between mother and embryo. The best known example of an transfer of information from the conceptus to the mother is the production of chorion gonadotrophin and sex steroids. As opposed to the classical concept that these hormones are first released after the initiation of implantation it has recently been postulated in a few somewhat controversial papers that both types of hormones are already produced *by the blastocyst* (see Lutwak-Mann, 1971; Noyes, 1972; Perry et al., 1973; Dickmann and Dey, 1974; Haour and Saxena, 1974; Dickmann et al., 1975; Sherman and Salomon, 1975; Sundaram et al., 1975). Local, non-systemic influences of the blastocyst on the directly surroundig maternal tissues have often been suspected, but seldom shown conclusively (with the exception of arylamidase: 3.2.2.2.1. and 4.4.1.; Denker, 1976c; van Hoorn and Denker, 1975).

A regulation of the process of *implantation* by *maternal hormones* has been conclusively demonstrated. If pregnant animals are ovariectomized shortly before implantation of the blastocysts, implantation will fail to occur in many species. In the rat and in the mouse deprivation of maternal estrogens so strongly reduces the metabolism of the blastocysts that their further development is largely stopped; they do not die, however, but remain in this "diapause stage" until estrogen is given. From this observation comes the concept of the initiation of implantation being triggered by an "*estrogen surge*" in the maternal blood (for a view of the comprehensive literature references see Psychoyos, 1973). In the rabbit and several other species (guinea pig, hamster, sheep: see Deanesly, 1960; Orsini and Psychoyos, 1965; Harper et al., 1966; Bindon, 1971a; Moore, 1975) such experiments have given no indication of a comparable role for maternal estrogens in the initiation of implantation. There is, however, in these species a clear dependence on ovarian *progesterone:* if, for example, rabbits are ovariectomized at 6 d p.c., the blastocysts do not implant nor enter a diapause stage, but die after a brief period of further expansion (Lutwak-Mann, et al., 1962). It can be shown that this absence of implantation corresponds to an absence in a comparable rise in the *blastolemmase activity* such as is observed in normal pregnancy (Denker, 1972). The blastocyst coverings are then not lysed and the trophoblast cannot establish contact with the uterine epithelium. If, however, a replacement therapy with progesterone is carried out, the blastolemmase activity is again normal and implantation goes along regularly.

These investigations lead us in the end to the conclusion that the hormonal regulation of the initiation of implantation probably occurs at least partly by a regulation of *implantation-initiating enzymes* (blastolemmase). On this basis alone we cannot decide, however, whether these ovarian hormones influence the enzyme activity directly or indirectly. On the basis of investigations mentioned done on the relationship of embryo to mother it is more probably correct to assume an *indirect* regulation, perhaps a change in the pattern of various essential components of the uterine secretion. In vitro culture experiments have usually shown that sex steroids have more an inhibiting than a supportive effect if adminstered directly to the embryo (for an outline of the literature see Beier, 1973, 1974b). Andary (1974), however, postulates a *direct* effect on the blastocyst of the mouse by sex steroids: here a chymotrypsin-like (GPNA-splitting) enzyme activity of the blastocyst is supposedly directly dependent on estradiol (or cAMP or prostaglandin E_2), which was concluded from a lack in the rise in the activity in in vitro culture of blastocysts when these ingredients were absent. The physiological role of this enzyme is unclear. A trypsin-like (BANA-splitting) enzyme activity of mouse blastocysts, which could have importance for the process of implantation judging from in-vivo experiments (see 4.4.2.3.), is not especially affected by this hormone in vitro, only a change in the compartmentalization in the blastocyst cells takes place. The postulated direct effect of the hormones on the blastocyst is not substantiated by the numerous physiological experiments mentioned above. As it is difficult to extrapolate from in vitro culture experiments to the situation in vivo, we must reserve our final judgement until more experiments have been done.

Inhibitors of the endometrium and uterine secretion could be of importance in the regulation of the process of implantation. Trypsin inhibitors have been demonstrated in the uterine secretion of the rabbit whose activity is dependent on the level of sex steroids in the maternal blood (Beier, 1970b). Apparently the entire female genital tract is rich in proteinase inhibitors (Blackwood et al., 1965; Schumacher, 1970; Schumacher and Zaneveld, 1974; Somerville and Dabich, 1974; Stambaugh et al., 1974; McLaughlin and Hamner, 1975; Wallner et al., 1976; regarding work done at the beginning of the century on the human endometrium see 4.4.2.3.). Even uteroglobin, a progesterone-dependent protein of low molecular weight which dominates in the uterine secretion of early pregnancy and pseudopregnancy in the rabbit, appears to possess trypsin inhibiting activity (Johnson, 1974; Beier, 1976). Apparently the blastolemmase activity is also influenced by such endometrial inhibitors (Denker, 1972). The trophoblast possibly contains large amounts of proteinase inhibitors, too, which could be a reason why the trophoblast as opposed to the highly active blastocyst coverings normally shows only a trace of blastolemmase activity in the substrate film test. If, however, trophoblast material is separated electrophoretically it is possible to demonstrate a very high blastolemmase activity (see 3.2.2.2.2.). Doubtless it would be biologically meaningful to have a high proteinase inhibitor activity at the site where blastolemmase (or a potent blastolemmase activator) is formed but where the enzyme cannot be allowed to become active.

We also have as an attractive hypothesis for the regulation of blastolemmase activity the possibility that the enzyme is first synthesized as a *proenzyme* and then transformed into the active state by other proteases (of the trophoblast, for example) acting as activators. This hypothesis awaits experimental scrutiny.

4.6. On the Role of the Trophoblast and the Endometrium at Implantation

Since the morphological studies done at the end of last century and mentioned in the introduction, but especially after the investigations done by Graf von Spee, the question regarding the active or passive role of the trophoblast and endometrium at implantation has arisen and been variously answered. Sometimes the embryo was compared to a malignant, invasive tumor; by others the embryo was regarded as relatively passive and it was pointed out that in some species (mouse) the uterine epithelium will disappear even around inert foreign bodies and oil drops, thereby triggering a decidual reaction to a certain extent (see Finn, 1971; McLaren and Nilsson, 1971; Hinchliffe and El Shershaby, 1975). In man, the important role of the functionally transformed endometrium has been emphasized (Schmidt-Matthiesen, 1967, 1968, 1970).

Probably it is wrong to ask about an active role for the trophoblast or the uterine mucosa in such a pointed manner. On one hand the *trophoblast* has a special role in the establishment of contact, for which it appears specialized. Embryonic knot cells are incapable of implantation (mouse: Gardner, 1972). The dissolution of the blastocyst coverings in the rabbit always begins where trophoblastic knobs lie next to them (see 3.1.1., 4.2., Fig. 2, 4, 28), while the coverings remain intact at first over the trophoblast ot the embryonic hemisphere who has no trophoblastic knobs, and over the embryonic disc, and are only mechanically pushed aside. The situation is similar in the cat (see 3.1.2.), the ferret (Enders and Schlafke, 1972), Citellus (Mossman and Weisfeldt, 1939) and the guinea pig (Spee, 1901; Blandau and Rumery, 1957). It was demonstrated in the rabbit that the lysis of the coverings begins at the abembryonic pole *independent from the orientation of the blastocyst in the uterus,* which means the coverings lysis will beginn at the trophoblastic pole even when it lies against atypical regions of the endometrium (see 3.3.1.2.). The blastolemmase activity always shows **a** maximum at the abembryonic pole. By a reversed orientation of the blastocys we often do find a remarkable proteinase activity, though, between the antimesometrial endometrium and the surface of the blastocyst (in this case the embryonic disc region), the meaning of which is not clear.

Embryonic coverings without a trophoblast, exposed to the uterine milieu, build an interesting model for a reduced system which is lacking in trophoblast-dependent factors (see 3.3.1.1.). These models do not develop the typical blastolemmase activity in the uterus and are only dissolved after great delay (see 3.3.1.1.). In degenerated rabbit eggs, the mucoprotein layer (but not the zona pellucida) remains intact in the uterus for a long time (see also Adams, 1970; Beier, 1973, 1974a and b). In the sheep the dissolution of the zona pellucida in unfertilized, degenerated embryos is strongly delayed (Bindon, 1969). All of these observations point to a major *role for the trophoblast* in the blastolemmase activity and the dissolution of the blastocyst coverings.

Chemical differences between trophoblast-dependent proteinase (blastolemmase) and uterine secretion endopeptidase have been established and were discussed above (4.4.2.1.). Proteinase inhibitors which inhibit blastolemmase hinder the dissolution of the coverings when applied in vivo (4.4.2.2.). This fits in well with the general picture we are developing of a special role for the trophoblast in the proteinase activity of the blastocysts and in implantation, but does not exclude a role for the uterine secretion proteinases as they are also inhibited by the inhibitors used. When antipain is applied, the abnormally enlarged trophoblastic knobs which are hindered in their implantation develop a high proteinase activity which, as opposed to the untreated controls, can be directly shown in histochemical tests. The morphological details of control blastocysts recognizable in semi-thin sections and under the electron microscope speak for an *erosion of the blastocyst coverings from the inside as well as from the outside.* After treatment with antipain or Trasylol®, however, the signs of erosion from the inside, from the trophoblastic knobs, predominate sometimes. This could be explained by a compensatorily enhanced production of proteinase (especially in the case of antipain) by the trophoblastic knobs (see 3.3.2.1., 4.4.2.2.).

On the other hand, the biochemistry of the *maternal tissues* doubtless plays an important role in the initiation of implantation. This is impressively demonstrated by the failure to implant caused by ovariectomy (see 4.5.). The uterine secretion contains various enzymes at the moment of implantation which can take part in the dissolution of the blastocyst coverings: in addition to the endopeptidase activity mentioned, we find glycosidases and a noticeably high amino acid arylamidase activity (see 4.4.1.). The release of this latter enzyme into the uterine lumen appears, interestingly enough, to be stimulated locally at the site of implantation by the blastocyst. The endopeptidase activity shown in gelatin substrate film tests is often especially high on the surface of those regions of the uterine epithelium to which the trophoblast is directly apposed; this is more noticeable in the cat than in the rabbit (see 3.2.2.2.2.). The possibility of an *activation of endometrial enzymes by the embryo undergoing implantation* has been especially emphasized in investigations done on other species (rodents, man, see 4.4.2.3.).

Thus we arrive at the concept that *maternal and embryonic activities work together meaningfully* in the complex molecular biological processes which take place during the attachment of the blastocyst to the endometrium. Doubtless the number of components which take part is greater than yet known and a completion of the actual picture necessitates more extensive investigations. From the described experiments it is clear that, at least in the rabbit, *proteinases* of the blastocyst as well as the endometrium) have a major function. The concept that the trophoblast-dependent proteinase activity (blastolemmase) develops its function at the interface between embryonic and maternal tissue, calls forth interesting *immunobiological problems.* If a mother animal carries allogeneic embryos, their implantation and development, even in repeated pregnancies, is similar to or even better than that of isogeneic conceptuses (see Beer et al., 1975). The basis of this phenomenon, which reminds one of the "enhancement" known from tumor and transplantation biology, is unknown. Are antibodies formed against trophoblast proteinases, and if so, do these neutralize or do they rather enhance enzyme activity (as has been shown in some cases for other enzymes)? Are the enzymes released as proenzymes? Which role do the inhibitors of the trophoblast and endometrium play, and can activators of endometrial enzymes be demonstrated in the trophoblast or v.v.? Hopefully further investigations will supply us with answers to these questions.

4.7. The Possible Use of Proteinase Inhibitors in the Inhibition of Implantation as a Method of Birth Control

The results of the experiments discussed here have shown that it is possible to very effectively inhibit the lysis of the blastocyst coverings in the rabbit by the application of proteinase inhibitors in vivo. The observed tardy and incomplete attachment of the trophoblast could perhaps also be blocked *if an adequately high intrauterine concentration of inhibitors can be maintained over a sufficient period of time.* The maintenance of a sufficently high intrauterine concentration of inhibitor would be realizable by "slow release devices" (Duncan and Kalkwarf, 1973). Perhaps inhibitors to be given orally could also be developed which would appear in high concentrations in the uterine secretion. Experiments whose goal is an effect on sperm acrosin are already being carried out in various laboratories (personal communications). Although substances such as TLCK or NPGB are excluded because of the possibility of side effects (see 4.4.2.2.1., 4.4.2.3) natural inhibitors such as Trasylol[®] which are well tolerated have found great interest. We saw no toxic *side effects* in the mother with these substances.

We also observed that the resorption of Trasylol® from the uterus takes place very slowly. With further knowledge of the specific properties of the trophoblastic enzyme (blastolemmase) a further selectivity for the pharmaceuticals is conceivable. This concept certainly deserves special attention because specific biochemical features of embryonic tissues are attacked here. This is opposed to those methods of contraception whose primary targets are maternal organs (ovary, tubes, uterus including cervix), for example the hormonal contraceptives.

For the present several important questions remain open which could only be answered by animal experiments which include other species: since with antipain treatment, for example, the embryonic disc undergoes early damage while the trophoblast continues growing (see 3.3.2.1., 4.4.2.2.2.), must we not consider the possibility of the occasional development of a vesicular mole? Should we fear an accumulation of implantations at abnormal, even *ectopic,* places since in our experiments the manipulation on the uterus in the late preimplantation phase (including the control injection) leads to an accumulation of cases of dystopic implantations?

Above all one must deal with the question whether or not it is at all desirable to base a method of contraception on interfering with any stage of embryonic development after the formation of the zygote.

Summary

Implantation in the rabbit and, for comparison, in the cat is used as a model to investigate processes which play a role in the *initiation of trophoblast attachment* to the uterine mucosa. In combined morphological, histochemical and biochemical experiments evidence is found that certain hydrolytic enzymes and especially *proteinases* of the blastocyst play an important role here; this concept was tested by applying specific proteinase inhibitors in vivo and it was shown that inactivation of the described proteinases effectively inhibits implantation.

The investigations are based on the *morphological* analysis of attachment in normal pregnancy of the cat and rabbit. Both species display coitus-induced ovulation, so that it is possible to determine the exact age of the embryo. At the beginning of implantation the blastocysts are relatively large (diameter ca. 5–6 mm), as both species represent the central type of implantation which greatly facilitates systematic morphological, histochemical and biochemical investigations. Morphological evidence for a lytic activity of the abembryonic trophoblast is found in both species.

In the *rabbit* when implantation is initiated the blastocyst is still surrounded by thick *coverings* of extracellular material, which mainly originated from the secretions of the tube, uterus and trophoblast. The lamellar composition of the blastocyst coverings is described morphologically and discussed; it is questionable whether or not in the rabbit after the strong expansion of the blastocyst a layer remains which could be identified as the zona pellucida. Implantation takes place in two different phases in this species, one abembryonic-antimesometrial (obplacentation, formation of a yolk sac placenta, beginning 7 d p.c.) and the other embryonic-mesometrial (formation of the definitive chorioallantoid placenta, beginning 8 d p.c.). Special attention is given to the introductory phase, the abembryonic attachment. At the initiation of implantation the blastocyst coverings are first dissolved at the abembryonic pole of the blastocyst by swelling and disappearance of their lamination, while they remain unchanged at the embryonic pole. The process of lysis begins over each symplasmatic element of the trophoblast, the *trophoblastic knobs.* The light microscopical and electron micros-copical morphology and behavior of these specialized areas of the trophoblast which represent the invasive elements are reported in detail. The trophoblastic knobs *fuse* selectively with those cells of the uterine cavum epithelium which sit atop a sub-epithelial capillary and their cytoplasm spreads out in the direction of the blood vessel and erodes it to establish a hemochorial contact. Until 7 d p.c. only a few cells of the cavum epithelium are fused together to form multinucleate complexes, while, in contrast to this, 7 1/2–8 d p.c. a wide-spread cell fusion with the *forma-tion of broad symplasms* takes place in the area around the blastocyst. This does not occur to any comparable extent in those regions between the sites of implanta-tion, indicating that it must he caused by a locally effective signal from the blasto-cyst whose (chemical?) nature is yet unknown.

In the *cat* the blastocysts are freely movable in the uterine lumen up to 13 d p.c. Up to 12 d p.c. the blastocyst coverings are at least partially intact. They are single-layered and can be referred to as the *zona pellucida,* which must be considerably swollen during the expansion of the blastocyst. As in the rabbit, the trophoblast pole of the blastocyst is the region where the coverings are first dissolved; in the region of the embryonic disc which in this species comes to lie antimesometrially, they remain intact longer. At 13 d p.c. they can no longer be detected. In the cat numerous leuco-cytes are found in the uterine lumen at these stages. The endometrial glands widen especially at the blastocyst site. The invasion of the trophoblast begins 14 d p.c. in the girdle-shaped region typical for carnivores.

Histochemical investigations of the *composition of the blastocyst coverings* of the rabbit show that they are comparable to epithelial mucins and possess protein com-ponents and carbohydrate groups with terminal neuraminic acid and sulfuric acid ester groups. Chemically related but slightly different mucosubstances form a *surface coat* on the uterine epithelium and, less prominently, at the surface of the trophoblast. The possible role of such glycoprotein-rich structures in cell adhesion is discussed. Models and possible pathways for the enzymatic degradation of blastocyst coverings and cell surface coats are discussed. These concepts are tested with experiments on the *dissolu-tion of the blastocyst coverings in vitro.*

Glycosidases and *proteases* which are expected to play a part here are histochemi-cally demonstrated in the trophoblast and endometrium, and the changes in their pattern of distribution and activity in the various phases of pregnancy and around the beginning of implantation are followed in the rabbit. In addition to the glycosidases (β-galactosidase, β-glucuronidase and β N-acetyl-glucosaminidase), we find considerable activity of *amino acid arylamidases* (exopeptidases of the aminopeptidase type) in the uterus and embryo. These arylamidases can be identified preliminarily on the basis of their substrate spectrum. One of the arylamidases (I) reaches a steep peak in activity in the uterine secretion around 5 d p.c. with values around 1000 mU/mg protein. Indica-tions were found in biochemical and histochemical tests that the *blastocysts* by an as yet unknown stimulus *locally enhance the release* of this enzyme from the uterine epithelium into the uterine secretion.

102

Special attention is given to a high *endopeptidase* (proteinase) activity of implanting blastocysts in the cat and the rabbit which can be detected by a histochemical gelatin substrate film test. The enzyme of rabbit blastocysts is shown by a large series of experiments with specific inhibitors to be a serine proteinase of the trypsin family. It has an interestingly narrow substrate spectrum and no activity worth mentioning with commonly used synthetic proteinase substrates. Its migratory behavior in micro disc electrophoresis and in agar gel electrophoresis is described and its differentiation from uterine secretion proteases including an enzyme which resembles trypsin more closely is discussed. The blastocyst enzyme is refered to as *blastolemmase*. Experiments with rabbit embryonic coverings lacking trophoblast tissue which were transplanted into pregnant or pseudopregnant uteri show clear evidence for a dependence of the enzyme on the (abembryonic) *trophoblast*. This is confirmed by the investigation of dystopically implanting embryos.

In the cat a particularly high activity of a *cathepsin*-B-like SH-proteinase of the trophoblast predominates. Since we could only carry out histochemical investigations on this material, we could not determine whether a blastolemmase-like enzyme is present which was being concealed by the high catheptic activity. High cathepsin-B-like activities could also be demonstrated in the entoderm and uterine stroma cells of the rabbit.

In attempts to *inhibit blastolemmase in vivo*, proteinase inhibitors which had proven very effective in vitro (Trasylol®, antipain, NPGB, SSPI) were applied intrauterally to rabbits at 6 1/2 d p.c., i. e. 1/2 day before the beginning of implantation. Blastolemmase activity is effectively inhibited by this treatment as is the lysis of the blastocyst coverings as followed in extensive morphological and histochemical tests. ϵ-Aminocaproic acid which does not inhibit blastolemmase was injected as a control and was found to be without any effect on implantation. Trasylol® proved itself especially effective. Determination of the concentration of inhibitor in the uterine flushings at different times after application showed that the elimination of Trasylol® is slower than that of antipain. NPGB toxically demages the blastocyst tissue, but electron microscopical investigations showed no signs of direct toxic effects of Trasylol® or antipain on the blastocyst tissues. The expansion of the treated blastocysts actually continues. The embryos remain surrounded by coverings after an effective dose of Trasylol® up to 7 1/2 d p.c.; after that they are mechanically disrupted by the continuing expansion of the blastocysts. Parts on the trophoblast which have been freed in such a manner can manage a late attachment at certain limited areas as soon as the intrauterine concentration of inhibitor has strongly decreased, i. e. 8 1/2 d p.c. This contact remains superficial and functionally insufficient, however. It is followed by grave disturbances in the uptake of maternal plasma proteins into the blastocyst cavity, and with increasing cell degeneration (which can first be detected in the region of the embryonic disc) the blastocyst dies and is resorbed.

These results are interpreted as support for the theory that *blastolemmase plays a special role in the dissolution of the blastocyst coverings and in the initiation of implantation.* The possible use of proteinase inhibitors to inhibit implantation in contraception is discussed and the problems involved are mentioned. The danger of an increase in the rate of dystopic implantation is emphasized as was observed in the process of our experiments with inhibitor injection as well as control injection.

Acknowledgements

I would like to thank Professor Dr. W. Kühnel for his continuing interest in my experiments and generous support of my work. I am indebted to Professor Dr. H. Fritz of the Abteilung für Klinische Chemie and Klinische Biochemie of the University of Munich for fruitful discussions and numerous suggestions on the subject of the biochemistry of proteinases and proteinase inhibitors as well as for the samples he contributed of the various inhibitors. I owe the cat material to the cooperation with Professor Dr. Ch. E. Hamner of the University of Virginia School of Medicine, Charlottesville, Virginia, USA. I would also like to thank my colleague Dr. Ursula Mootz for her friendly cooperation with the laparotomies and Professor Dr. Chr. Stang-Voss for her help with taking the electron micrographs. I thank Professor Dr. Dr. H. M. Beier for many stimulating discussions. The micro-disc electrophoreses are a result of work done with Dr. U. Petzoldt, formerly a member of the Arbeitsgruppe Professor Dr. G. H. M. Gottschewski at the Max Planck Institut für Immunbiologie, Freiburg, presently at the University College, London. The biochemical aminopeptidase tests were largely carried out by Miss Gerhild van Hoorn, working as a graduate student in our Freiburg group. I would like to thank Dr. E. Truscheit and Dr. W. Wingender, Bayer-AG, Wuppertal-Elberfeld, for the samples of Trasylol® and antipain and Prof. Dr. H. G. Schwick and Dr. N. Heimburger, Behring-werke, Marburg, for the samples of α_1-antitrypsin and α_1-antichymotrypsin.

I would particularly like to voice my thanks to the many members of our department who helped me carry out the experiments and who did the typing: Mrs. Ingeborg Ackermann, Mrs. Ria Becht, Miss Gerda Bohr, Mrs. Sabine Flick, Miss Edith Höricht, Mrs. Jutta Jacobs, Miss Petra Moeller, Mrs. Maria Petuelli and Miss Elke Schmale, as well as the photographer Mr. G. Plitzner, and the graphic artist Mr. W. Graulich.

The experiments were generously supported by the Deutsche Forschungsgemeinschaft as part of the Schwerpunktprogramm "Physiologie und Pathologie der Fortpflanzung" (De 181/3 and 4).

References

Abe, M.: Investigation of placental ferments in various stages of pregnancy. Jap. J. Obstet. Gynec. **15**, 44–52 (1932)

Abraham, R., Fulfs, J. C., Golberg, L., Coulston, F.: Cytotoxic action of a mixed copolymer of phenylmethylcyclosiloxane on rabbit blastocyst lysosomes. J. Reprod. Fertil. **34**, 451–456 (1973)

Abraham, R., Hendy, R., Dougherty, W. J., Fulfs, J. C., Golberg, L.: Participation of lysosomes in early implantation in the rabbit. Exp. mol. Path. **13**, 329–345 (1970)

Adams, C. E.: The fate of unfertilized eggs in the rabbit. J. Reprod. Fertil. **23**, 319–324 (1970)

Adams, C. E.: The reproductive status of female mink, Mustela vison, recorded as failed to mate. J. Reprod. Fertil. **33**, 525–529 (1973)

Albers, H. J., Bedford, J. M., Chang, M. C.: Uterine peptidase activity in the rat and rabbit during pseudopregnancy. Amer. J. Physiol. **201**, 554–556 (1961)

Alloiteau, J. J., Psychoyos, A.: "Y a-t-il pour l'oeuf de ratte deux façons de perdre sa zone pellucide?" C. R. Acad. Sci. (Paris) Sér. D **262**, 1561–1564 (1966)

Amoroso, E. C.: Placentation. In: Marshall's Physiology of Reproduction. A. S. Parkes (ed.), pp. 127–311. London: Longmans, Green & Co. 1961

Amoroso, E. C.: In Diskussion. In: Ciba Foundation Study Group No. 23 on Egg Implantation. G. E. W. Wolstenholme and M. O'Connor (eds.), p. 100. London: J & A. Churchill, 1966

Andary, T. J.: Trypsin – and chymotrypsin – like enzymes in preimplantation mouse embryos. Diss. Wayne State Univ. Detroit, USA 1974

Andary, T. J., Dabich, D.: A sensitive polyacrylamide disc gel method for detection of proteinases. Analyt. Biochem. **57**, 457–466 (1974)

Andary, T. J., Dabich, D., van Winkle, L. J.: Changes in proteinase activity in early vs. late mouse blastocysts. J. Cell Biol. **55**, 3a (1972)

Andersson, L., Nilsson, I. M., Nilehn, J. E., Hedner, U., Grandstrand, B., Melander, B.: Experimental and clinical studies on AMCA, the antifibrinolytically active isomer of p-aminomethyl cyclohexane carboxylic acid. Scand. J. Haemat. **2**, 230–247 (1965)

Anton, E., Brandes, D., Barnard, S.: Lysosomes in uterine involution. Distribution of acid hydrolases in luminal epithelium. Anat. Rec. **164**, 231–252 (1969)

Assheton, R.: On the causes which lead to the attachment of the mammalian embryo o the walls of the uterus. Quart. J. micr. Sci. **97**, 173–190 (1895)

Bacsich, P., Hamilton, W. J.: Some observations on vitally stained rabbit ova with special references to their albuminous coat. J. Embryol. exp. Morph. **2**, 81–86 (1954)

Bang, N. U.: Physiology and biochemistry of fibrinolysis. In: Thrombosis and Bleeding Disorders. Theory and Methods. N. U. Bang, F. K. Beller, E. Deutsch and E. F. Mammen (eds.), pp. 292–327. Stuttgart: Thieme; New York, London: Academic Press 1971

Bang, N. U., Beller, F. K., Deutsch, E., Mammen, E. F.: Thrombosis and Bleeding Disorders. Theory and Methods. Stuttgart: Thieme; New York, London: Academic Press, 1971

Barrau, M. D., Abel jr., J. H., Verhage, H. G., Tietz jr., W. J.: Development of the endometrium during the estrous cycle in the bitch. Amer. J. Anat. **142**, 47–65 (1975a)

Barrau, M. D., Abel jr., J. H., Torbit, C. A., Tietz jr., W. J.: Development of the implantation chamber in the pregnant bitch. Amer. J. Anat. **143**, 115–130 (1975b)

Barrett, A. J.: Inhibitors of lysosomal proteinases. In: Bayer-Symposium V: Proteinase Inhibitors. H. Fritz, H. Tschesche, L. J. Greene, E. Truscheit, (eds.), pp. 574–580. Berlin, Heidelberg, New York: Springer-Verlag 1974

Barrett, D.: Hatching enzyme of the sea urchin, Strongylocentrotus purpuratus. Amer. Zoologist **8**, 816–817 (1968)

Barrett, D., Angelo, G. M.: Maternal characteristics of hatching enzymes in hybrid sea urchin embryos. Exp. Cell. Res. **57**, 159–166 (1969)

Beckman, L., Björling, G., Christodoulou, C.: Pregnancy enzymes and placental polymorphism: II. Leucine aminopeptidase. Acta genet. (Basel) **16**, 122–131 (1966)

Beer, A. E., Billingham, R. E., Scott, J. R.: Immunogenetic aspects of implantation, placentation and feto-placental growth rates. Biol. of Reprod. **12**, 176–189 (1975)

Beier, H. M.: Biochemisch-entwicklungsphysiologische Untersuchungen am Proteinmilieu für die Blastozystenentwicklung des Kaninchens (Oryctolagus cuniculus). Zool. Jahrb. Abt. Anat. Ontog. Tiere **85**, 72–190 (1968a)

Beier, H. M.: Uteroglobin: A hormone – sensitive endometrial protein involved in blastocyst development. Biochim. Biophys. Acta (Amst.) **160**, 289–291 (1968b)

Beier, H. M.: Protein patterns of endometrial secretion in the rabbit. In: Ovo-Implantation. Human Gonadotropins and Prolactin. Second Seminar on Reproductive Physiol. and Sexual Endocrinol. Brussels 1968. P. O. Hubinot et al. (eds). pp. 157–163. Basel, München, New York: Karger 1970a

Beier, H. M.: Hormonal stimulation of protease inhibitor activity in endometrial secretion during early pregnancy. Acta endoc. **63**, 141–149 (1970b)

Beier, H. M.: Die Pseudogravidität des Kaninchens nach Stimulierung mit Choriongonadotropin. Diss. Fachbereich Humanmedizin, Marburg 1971

Beier, H. M.: Die hormonelle Steuerung der Uterussekretion und frühen Embryonalentwicklung des Kaninchens. Habilitationsschrift Univ. Kiel 1973

Beier, H. M.: Oviducal and uterine fluids. J. Reprod. Fertil. **37**, 221–237 (1974a)

Beier, H. M.: Ovarian steroids in embryonic development before nidation. In: Hormones and Embryonic Development. G. Raspé (ed.), pp. 199–219. Advanced in the Biosciences **13**, Oxford: Pergamon Press; Braunschweig: Friedr. Vieweg & Sohn, 1974b

Beier, H. M.: Uteroglobin and related biochemical changes in the reproductive tract during early pregnancy in the rabbit. J. Reprod. Fertil., Suppl. **25**, 53–69 (1976)

Beier, H. M., Kühnel, W.: Die verzögerte Uterussekretion nach Östrogeninjektionen beim graviden Kaninchen. Verh. anat. Ges. (Jena) 67, Anat. Anz. Suppl. **134**, 567–575 (1973)

Beier, H. M., Kühnel, W., Petry, G.: Uterine secretion proteins as extrinsic factors in preimplantation development. Schering Symp. on Intrinsic and Extrinsic Factors in Early Mamm. Devel., Venice 1970. G. Raspé (ed.), pp. 165–189. Advances in the Biosciences 6. Oxford: Pergamon Press; Braunschweig: Friedr. Vieweg & Sohn, 1971

Beier, H. M., Kühnel, W., Petry, G.: Morphologische und biochemische Befunde am pseudograviden Kaninchenendometrium nach gonadotroper Stimulierung. Verh. anat. Ges. (Jena) **66**, Anat. Anz. Suppl. **130**, 445–457 (1972a)

Beier, H. M., Mootz, U., Kühnel, W.: Asynchrone Eitransplantationen während der verzögerten Uterussekretion beim Kaninchen. VII. Internat. Kongr. f. Tierische Fortpflanzung und Haustierbesamung, München. pp. 1891–1896, 1972b

Beneden van, E., Julin, C.: Recherches sur la formation des annexes foetales chez les mammifères: Lapin et Cheiroptères. Arch. Biol. **5**, 369–434 (1884)

Bennett, M. V. L.: Function of electronic junctions in embryonic and adult tissue. Fed. Proc. **32**, 65–75 (1973)

Bergmeyer, H. U.: Methoden der enzymatischen Analyse. 3. neubearb. und erw. Aufl. Weinheim/Bergstr.: Verlag Chemie 1974

Bergström, S.: Estimation of proteolytic activity at mouse implantation sites by the gelatin digestion method. J. Reprod. Fertil. **23**, 481–485 (1970)

Bergström, S.: Surface ultrastructure of mouse blastocysts before and at implantation Thesis, Uppsala 1971

Bergström, S.: Shedding of the zona pellucida in normal pregnancy and in various hormonal states in the mouse. A scanning electron microscope study. Z. Anat. Entwickl.-Gesch. **136**, 143–167 (1972a)

Bergström, S.: Histochemical localization of acid uterine aminoacylnaphthylamidases in early pregnancy and in different hormonal states of the mouse. J. Reprod. Fertil. **30**, 177–183 (1972b)

Bergström, S., Lutwak-Mann, C.: Surface ultrastructure of the rabbit blastocyst. J. Reprod. Fertil. **36**, 421–422 (1974)

Bhargava, A. S., Buddecke, E., Werries, E., Gottschalk, A.: Studies on glycoproteins, XIV. O-seryl-N-acetylgalactosaminide glycosidase, the enzyme splitting the O-glycosidic linkage between carbohydrate and peptide in bovine submaxillary glycoprotein. Biochim. Biophys. Acta **127**, 457–467 (1966)

Biggers, J. D., Stern, S.: Metabolism of the preimplantation mammalian embryo. In: Advances in Reprod. Physiol. M. W. H. Bishop (ed.), 6, pp. 1–59, London: Paul Elek Ltd., 1973

Billington, W. D.: Biology of the trophoblast. In: Advances in Reprod. Physiol. M. W. H. Bishop (ed.) **5**, pp. 27–66. London: Logos Press, 1971

106

Bindon, B. M.: Fate of the unfertilized sheep ovum. J. Reprod. Fertil. **20**, 183–184 (1969)

Bindon, B. M.: Role of progesterone in implantation in the sheep. J. Reprod. Fertil **24**, 146 (1971a)

Bindon, B. M.: Systematic study of preimplantation stages of pregnancy in the sheep. Austral. J. Biol. Sci. **24**, 131–147 (1971b)

Bitton-Casimiri, K., Brun, J.-L., Psychoyos, A.: Comportement in vitro des blastocysts du 5e jour de la gestation chez la Ratte; étude micro-cinématographique. C. R. Acad. Sci. (Paris) Sér. D. **270**, 2979–2982 (1970)

Blackwood, C. E., Hosannah, Y., Mandl, I.: Proteolytic enzyme systems in developing rat tissues. J. Reprod. Fertil.**17**, 19–33 (1968)

Blackwood, C. E., Mandl, I., Long, M. E.: Proteolytic enzymes and their inhibitors in human gynecological tumors. Am. J. Obstet. Gynecol. **91**, 419–429 (1965)

Blandau, R. J.: Embryo-endometrial interrelationship in the rat and guinea pig. Anat. Rec. **104**, 331–359 (1949)

Blandau, R. J.(ed.): The Biology of the Blastocyst. Chicago, London: Univ. of Chicago Press 1971a

Blandau, R. J.: Culture of guinea pig plastocyst. In: The Biology of the Blastocyst. R. J. Blandau (ed.), pp. 59–69. Chicago, London: Univ. of Chicago Press, 1971b

Blandau, R. J., Rumery, R. E.: The attachment cone of the guinea-pig blastocyst as observed under time-lapse cinematography. Fertil. and Steril. **8**, 570–585 (1957)

Böving, B. G.: Internal observation of rabbit uterus. Science **116**, 211–214 (1952)

Böving, B. G.: Blastocyst-uterine relationships. In: Cold Spring Harbor Symp. Quant. Biol. **19**, 9–28 (1954)

Böving, B. G.: Rabbit egg coverings. Anat. Rec. **127**, 270 (1957)

Böving, B. G.: Implantation. Ann. New York Acad. Sci. **75**, 700–725 (1959)

Böving, B. G.: L'interaction entre les méchanismes physiologiques intervenant dans l'implantation du blastocyste chez la lapine. In: Les Fonctions de Nidation Uterine et leur Troubles. Colloque de la Sociéte Nationale pour l'Étude de la Stérilité et de la Fécondité. Rapporteurs generaux: J. Ferin, M. Gaudefroy. pp. 103–124. Paris: Masson et Cie. 1960

Böving, B. G.: Anatomical analysis of rabbit trophoblast invasion. Carnegie Inst. Washington, Publ. No. 621, Contrib. Embryol. **37**, No. 254, 33–55 (1962)

Böving, B. G.: Implantation mechanisms. In: Conference on Physiological Mechanisms Concerned with Conception. C. G. Hartman (ed.), pp. 321–396. Oxford, London, New York, Paris: Pergamon Press, 1963

Böving, B. G.: Anatomy of Reproduction. In: Obstetrics. J. F. Greenhill (ed.), 13th edition, second printing pp, 3–102. Philadelphia: Saunders 1970

Böving, B. G., Larsen, J. F.: Implantation. In: Human Reproduction. Conception and Contraception. E. S. E. Hafez, T. N. Evans (eds.), pp. 133–156. Hagerstown (Maryland), Med. Dept. Harper & Row, Publ., 1973

Bowman, P., McLaren, A.: The reaction of the mouse blastocyst and its zona pellucida to enzymes in vitro. J. Embryol. exp. Morph. **24**, 331–334 (1970)

Bradbury, S., Billington, W. D., the late Kirby, D. R. S., Williams, E. A.: Histochemical characterization of the surface mucoprotein of normal and abnormal human trophoblast. Histochem. J. **2**, 263–274 (1970)

Brambell, F. W. R., Hemmings, W. A.: The passage into the embryonic yolk-sac cavity of plasma proteins in the rabbit. J. Physiol. (Lond.) **108**, 177–184 (1949)

Brun, J. L., Psychoyos, A.: Dissolution of the rat zona pellucida by acidified media and blastocyst viability. J. Reprod. Fertil. **30**, 489–491 (1972)

Buddecke, E., Schauer, H., Werries, E., Gottschalk, A.: Characterization of 0-seryl-N-acetylgalactosaminide glycohydrolase as an α-N-acetylgalactosaminidase. Biochem. Biophys. Res. Commun. **34**, 517–521 (1969)

Busch, L., Mootz, U., Kühnel, W.: Zur Oberflächenbeschaffenheit der Schleimhaut von Tube und Uterus des Kaninchens im Oestrus. Verh. Anat. Ges. **71**, 525–530 (1977)

Caffier, P.: Die proteolytische Fähigkeit von Ei und Eibett. Zbl. Gynäk. **53**, 1910–1911 (1929a)

Caffier, P.: Die proteolytische Fähigkeit von Ei und Eibett (Experimentelle Studien mit Chorion und Dezidua). Zbl. Gynäk. **53**, 2410–2425 (1929b)

Caravita, S., Zacchei, A.M.: The anionic binding-sites at the cell surface after tissue dissociation and during the early phase of cell reaggregation. J. Embryol. exp. Morph. **32**, 35–55 (1974)

Carol, W., Klinger, G., Hempel, E.: Zur Möglichkeit einer postkonzeptionellen Fertilitatskontrolle durch medikamentöse Implantationshemmung. Zbl. Gynäk. **95**, 1761-1767 (1973)

Carroll, E. J., Epel, D.: Isolation and biochemical activity of the proteases released by sea urchin eggs following fertilization. Develop. Biol. **44**, 22–32 (1975)

Carroll, E. J. jr., Hedrick, J. L.: Hatching in the toad Xenopus laevis: Morphological events and evidence for a hatching enzyme. Develop. Biol. **38**, 1–13 (1974)

Chang, M. C.: Development and fate of transferred rabbit ova or blastocyst in relation to the ovulation time of recipients. J. exp. Zool. **114**, 197–225 (1950)

Chang, M. C., Hunt, D. M.: Effects of proteolytic enzymes on the zona pellucida of fertilized and unfertilized mammalian eggs. Exp. Cell Res. **11**, 497–499 (1956)

Chase, T. jr., Shaw, E.: Titration of trypsin, plasmin, and thrombin with p-nitrophenyl-p'-guanidino-benzoate HCl. In: Methods in Enzymology, S. P. Colowick and N. O. Kaplan (eds.) XIX: Proteolytic Enzymes. G. E. Perlman and L. Lorand (Vol.-eds.), pp. 20–27. New York, London: Academic Press 1970

Christie, G. A.: Comparative histochemical distribution of "leucine aminopeptidase" in the placenta and foetal membranes. Histochemie **10**, 272–277 (1967)

Cole, R. J.: Cinematographic observations on the trophoblast and zona pellucida of the mouse blastocyst. J. Embryol. exp. Morph. **17**, 481–490 (1967)

Coleman, R. L., Scroggs, R. A., Whittington, A.: Purification and properties of β-N-acetylglucos-aminidase from bovine uterus. Biochim. Biophys. Acta **146**, 290–292 (1967)

Conchie, J., Findlay, J.: Influence of gonadectomy, sex hormones and other factors on the activity of certain glycosidases in the rat and mouse. J. Endocrinol. **18**, 132–146 (1959)

Conchie, J., Strachan, I.: Distribution, purification and properties of 1-aspartamido-β-N-acetyl-glucosamine amidohydrolase. Biochem. J. **115**, 709–715 (1969)

Conrad, K., Buckley, J., Stambaugh, R.: Studies on the nature of the block to polyspermy in rabbit ova. J. Reprod. Fertil. **27**, 133–135 (1971)

Cook, G. M. W., Stoddart, R. W.: Surface carbohydrates of the eukaryotic cell. London, New York: Academic Press, 1973

Courrier, R., Gros, G.: Contribution a l'étude du cycle génital chez la Chatte. C. R. Soc. Biol. Filiales (Paris) **110**, 51–53 (1932a)

Courrier, R., Gros, G.: Remarques sur la nidation de l'oeuf chez la Chatte. C. R. Soc. Bjol. Filiales (Paris) **111**, 787–789 (1932b)

Courrier, R., Gros, G.: Données complémentaires sur le cycle génital de la Chatte. C. R. Soc. Biol. Filiales (Paris) **114**, 275–277 (1933)

Culling, C. F. A., Reid, P. E., Clay, M. G., Dunn, U. L.: The histochemical demonstration of 0-acyl-ated sialic acid in gastrointestinal mucins. Their asscociation with the potassium hydroxyde-periodic acid-Schiff-effect. J. Histochem. Cytochem. **22**, 826–831 (1974)

Curtis, A. S. G.: The cell surface: its molecular role in morphogenesis. London, New York: Logos, Press, Academic Press 1967

Curtis, A. S. G.: On the occurrence of specific adhesion between cells. J. Embryol. exp. Morph. **23**, 253–272 (1970)

Dabich, D., Andary, T. J.: Prevention of blastocyst implantation in mice with proteinase inhibitors. Fertil. and Steril. **25**, 954–957 (1974)

Dabich, D., Andary, T. J.: Tryptic- and chymotryptic-like proteinases in early and late preimplantation mouse blastocysts. Biochim. Biophys. Acta **444**, 147–153 (1976)

Dallenbach-Hellweg, G.: Histopathology of the Endometrium. Engl. translation by F. D. Dallenbach. Second revised and enlarged edition. Berlin, Heidelberg, New York: Springer-Verlag 1975

Davies, J., Hoffman, L. H.: Studies on the progestational endometrium of the rabbit. I. Light microscopy, day 0 to day 13 of gonadotrophin-induced pseudopregnancy. Amer. J. Anat. **137**, 423–446 (1973)

Davies, J., Hoffman, L. H.: Studies on the progestational endometrium of the rabbit. II. Electron microscopy, day 0 to day 13 of gonadotrophin-induced pseudopregnany. Amer. J. Anat. **142**, 335–366 (1975)

Dawson, A. B., Kosters, B. A.: Preimplantation changes in the uterine mucosa of the cat. Amer. J. Anat. **75**, 1–37 (1944)

Deanesly, R.: Implantation and early pregnancy in ovariectomized guinea-pigs. J. Reprod. Fertil. **1**, 242–248 (1960)

Denker, H.-W.: Zur Enzym-Topochemie von Frühentwicklung und Implantation des Kaninchens. Dissertation, Med. Fak. Marburg 1969

Denker, H.-W.: Topochemie hochmolekularer Kohlenhydratsubstanzen in Frühentwicklung und Implantation des Kaninchens. I. Allgemeine Lokalisierung und Charakterisierung hochmolekularer Kohlenhydratsubstanzen in frühen Embryonalstadien. Zool. Jahrb. Abt. Allgem. Zool. und Physiol. 75, 141–245 (1970a)

Denker, H.–W.: Topochemie hochmolekularer Kohlenhydratsubstanzen in Frühentwicklung und Implantation des Kaninchens. II. Beiträge zu entwicklungsphysiologischen Fragestellungen. Zool. Jahrb. Abt. Allgem. Zool. und Physiol. 75, 246–308 (1970b)

Denker, H.-W.: Artefakte im Amylase-Substratfilmtest. Histochemie 21, 17–20 (1970c)

Denker, H.-W.: Enzym-Topochemie von Frühentwicklung und Implantation des Kaninchens. I. Glykogenstoffwechsel. Histochemie 25, 256–267 (1971a)

Denker, H.-W.: Enzym-Topochemie von Frühentwicklung und Implantation des Kaninchens. II. Glykosidasen. Histochemie 25, 268–285 (1971b)

Denker, H.-W.: Enzym-Topochemie von Frühentwicklung und Implantation des Kaninchens. III. Proteasen. Histochemie 25, 344–360 (1971c)

Denker, H.-W.: Substratfilmtest für den Proteasennachweis. XIII. Sympos. Ges. f. Histochemie, Graz 1969. Acta Histochem. Suppl. (Jena) X, 303–305 (1971d)

Denker, H.-W.: Blastocyst protease and implantation: Effect of ovariectomy and progesterone substitution in the rabbit. Acta endocrinol. (Kbh.) 70, 591–602 (1972)

Denker, H.-W.: Blastocyst enzymes and implantation in the rabbit. Abstract, VIth Ann. Meet. Soc. Study of Reproduct., Athens, Georgia, 1973. Biol. Reproduct. 9, 102–103 (1973)

Denker, H.-W.: Protease substrate film test. Histochemistry 38, 331–338 and 39, 193 (1974a)

Denker, H.-W.: Trophoblastic factors involved in lysis of the blastocyst coverings and in implantation in the rabbit: Observations on inversely orientated blastocysts. J. Embryol. exp. Morph. 32, 739–748 (1974b)

Denker, H.-W.: Implantation des Kaninchenkeims im Uterus: Bedeutung von Enzymen des Embryos bzw. des Endometriums. Verh. Anat. Ges. 69, 281–289 (1975)

Denker, H.-W.: Interaction of proteinase inhibitors with blastocyst proteinases involved in implantation. In: Protides of the Biological Fluids. Proceedings of the XXIIIrd Colloquium, Brugge 1975. H. Peeters (ed.), pp. 63–68. Oxford: Pergamon Press, 1976a

Denker, H.-W.: Wechselbeziehungen zwischen Blastozyste und Endometrium bei der Implantation: Beeinflussung der endometrialen Aminosäure-Arylamidase-Aktivität durch die Blastozyste. Verh. Anat. Ges. 70, Anat. Anz. Suppl. 140, 839–847 (1976b)

Denker, H.-W.: Effects of blastocysts and of copper IUDs on endometrial arylamidase in the rabbit. Anatomical Society of Great Britain and Ireland, Spring Meeting, Cambridge 1976. J. Anat. (Lond.) 122, 720 (1976c)

Denker, H.-W.: Copper IUD-induced loss of endometrial arylamidase activity in the rabbit. Biol. Reprod. 15, 519–522 (1976d)

Denker, H.-W.: Zur Spezifität und Empfindlichkeit des Proteasen-Substratfilmtests auf Gelatinebasis. XVIII. Sympos. Ges. f. Histochemie, Bozen 1975. Acta Histochem. Suppl. XVIII, 153–158 (1977)

Denker, H.-W., Hafez, E. S. E.: Proteases and implantation in the rabbit: Role of trophoblast vs. uterine secretion. Cytobiologie 11, 101–109 (1975)

Denker, H.-W., Hamner, C. E., Eng, L. A.: Studies on the early development and implantation in the cat. II. Implantation: proteinases. (in preparation)

Denker, H.-W., Hamner, C. E., Mootz, U., Eng, L. A.: Studies on the early development and implantation in the cat. I. Cleavage and blastocyst formation: differentiation of trophoblast and embryonic knot cells (in preparation)

Denker, H.-W., van Hoorn, G.: Peptidases related to implantation in the rabbit: Local stimulation of endometrial arylamidase secretion by the blastocyst immediately preceding implantation. Abstract, VIIth Ann. Meet. Society for the Study of Reproduction, Ottawa, Canada (1974)

Denker, H.-W., Kühnel, W.: Experimente zum Wirkungsmechanismus von Kupfer-IUDs beim Kaninchen: Lokale Beeinflussung der endometrialen Aminosäure-Arylamidase-Aktivität durch Kupfer-IUDs. Verh. Anat. Ges. 71, Anat. Anz. Suppl. 142, 531–536 (1977)

Denker, H.-W., Petzoldt, U.: Proteinases involved in implantation initiation in the rabbit: microdisc electrophoretic studies. Cytobiologie 15, 363–371 (1977)

Denker, H.-W., Stangl, R.: Versuche zur Lokalisierung und Abgrenzung verschiedener Aminosäure-Arylamidasen in Uterus und Blastozyste des Kaninchens. XVII. Sympos. Ges. f. Histochemie, Bozen 1974. Acta Histochem. Suppl. **XVI**, 249–257 (1976)

Dickmann, Z.: Shedding of the zona pellucida. In: Advances in Reprod. Physiol. A. McLaren (ed.), **4**, pp. 187–205. London: Logos Press 1969

Dickmann, Z., de Feo, V. J.: The rat blastocyst during normal pregancy and during delayed implantation, including an observation on the shedding of the zona pellucida. J. Reprod. Fertil. **13**, 3–9 (1967)

Dickmann, Z., Dey, S. K.: Steroidogenesis in the preimplantation rat embryo and its possible influence on morula-blastocyst transformation and implantation. J. Reprod. Fertil. **37**, 91–93 (1974)

Dickmann, Z., Dey, S. K., Gupta, J. S.: Steroidogenesis in rabbit preimplantation embryos. Proc. nat. Acad. Sci. (Wash.) **72**, 298–300 (1975)

Doorman, J. D.: De vasthechting van de kiemblaas aan den uteruswand bij het konijn. Diss. Univ. of Leiden Utrecht 1893

Duncan, G. W., Kalkwarf, D. R.: Sustained release systems for fertility control. In: Human Reproduction. Conception and Contraception. E. S. E. Hafez and T. N. Evans (eds.), pp. 483–504. Hagerstown (Md.), New York: Harper & Row, Publ. 1973

Duncan, G. W., Wheeler, R. G.: Pharmacological and mechanical control of implantation. Biol. of Reprod. **12**, 143–175 (1975)

Duval, M.: Le placenta des rongeurs. J. Anat. Physiol. (Paris) **25**, 309–342 (1889a)

Duval, M.: Le placenta des rongeurs: Le placenta du lapin. J. Anat. Physiol. (Paris) **25**, 573–627 (1889b)

Duval, M.: Le placenta des rongeurs: Le placenta du lapin (Suite). J. Anat. Physiol. (Paris) **26**, 1–48 et 273–344 (1890)

Enders, A. C.: The fine structure of the blastocyst. In: The Biology of the Blastocyst. R. J. Blandau (ed.), pp. 71–94. Chicago, London: Univ. of Chicago Press, 1971

Enders, A. C., Schlafke, S.: A morphological analysis of the early implantation stages in the rat. Amer. J. Anat. **120**, 185–226 (1967)

Enders, A. C., Schlafke, S.: Cytological aspects of trophoblast-uterine interaction in early implantation. Amer. J. Anat. **125**, 1–30 (1969)

Enders, A. C., Schlafke, S.: Penetration of the uterine epithelium during implantation in the rabbit. Amer. J. Anat. **132**, 219–240 (1971)

Enders, A. C., Schlafke, S.: Implantation in the ferret: Epithelial penetration. Amer. J. Anat. **133**, 291–316 (1972)

Enders, A. C., Schlafke, S.: Surface coats of the mouse blastocyst and uterus during the preimplantation period. Anat. Rec. **180**, 31–46 (1974)

Faillard, H., Pribilla, W.: Untersuchungen zur physiologischen Bedeutung der Neuraminsäure-haltigen Kohlenhydratgruppe von „Intrinsic-Factor"-Mucoiden menschlicher Magenschleimhaut. Klin. Wschr. **42**, 686–693 (1964)

Finn, C. A.: The biology of decidual cells. In: Adv. Reprod. Physiol. M. W. H. Bishop (ed), **5**, pp. 1–26 London: Logos Press 1971

Finn, C. A., Porter, D. G.: The uterus. London: Paul Elek (Scientific Books) Ltd., 1975

Fishman, W. H., Fishman, L. W.: The elevation of uterine β-glucuronidase activity by estrogenic hormones. J. Biol. Chem. **152**, 487–488 (1944)

Fox, L. L., Shivers, C. A.: Immunologic evidence for addition of oviductal components to the hamster zona pellucida. Fertil. Steril. **26**, 599–608 (1975)

Fraenkel, L.: Die Funktion des Corpus luteum. Arch. Gynäk. **68**, 438 (1903)

Frankl, O., Aschner, B.: Zur quantitativen Destimmung des tryptischen Fermentes in der Uterusmukosa. Gynäk. Rundsch. (Wien) **5**, 647–654 (1911)

Fritz, H., Hochstrasser, K.: Proteinase (elastase) inhibitors from dog submandibular glands. In: Methods in Enzymology. S. P. Colowick and N. O. Kaplan (eds.). Vol. XLV: Proteolytic Enzymes, Part B. L. Lorand (Vol.-ed.), pp. 860–869. New York, San Francisco, London: Academic Press 1976

Fritz, H., Jaumann, E., Meister, R., Pasquay, P., Hochstrasser, K., Fink, E.: Proteinase inhibitors from dog submandibular glands: isolation, amino acid composition, inhibition spectrum. In: Proceed. Internat. Research Conf. on Proteinase Inhibitors. H. Fritz and H. Tschesche (eds.), pp. 257–270. Berlin, New York: Walter de Gruyter 1971

110

Fritz, H., Schleuning, W. D., Schill, W.-B.: Biochemistry and clinical significance of the trypsin-like proteinase acrosin from boar and human spermatozoa. In: Bayer-Symp. V: Proteinase Inhibitors. Proceedings of the 2nd Internat. Research Conf. H. Fritz, H. Tschesche, L. J. Greene, and E. Truscheit (eds.), pp. 118–127. Berlin, Heidelberg, New York: Springer-Verlag, 1974a

Fritz, H., Trautschold, I., Werle, E.: Protease-Inhibitoren. In: Methoden der enzymatischen Analyse. H. U. Bergmeyer (ed.), pp. 1021–1038. 3. Aufl. Weinheim/Bergstr.: Verlag Chemie 1974b

Fritz, H., Schiessler, H., Schill, W.-B., Tschesche, H., Heimburger, N., Wallner, O.: Low molecular weight proteinase (acrosin) inhibitors from human and boar seminal plasma and spermatozoa and human cervical mucus: isolation, properties and biological aspects. In: Proteases and Biological Control. E. Reich, D. Rifkin, and E. Shaw (eds.), pp. 737–766. Cold Spring Harbor Laboratory, 1975a

Fritz, H., Schleuning, W.-D., Schiessler, H., Schill, W.-B., Wendt, V., Winkler, G.: Boar, bull and human sperm acrosin: isolation, properties and biological aspects. In: Proteases and Biological Control. E. Reich, D. Rifkin, and E. Shaw (eds.), pp. 715–735. Cold Spring Harbor Laboratory, 1975b

Fritz, H., Schleuning, W.-D., Schiessler, H., Schill, W.-B., Wendt, V.: Seminal proteinase inhibitors and sperm acrosin: Biochemistry and possible function. In: Human Semen and Fertility Regulation in the Male. E. S. E. Hafez (ed.), pp. 201–216. Saint Louis: C. V. Mosby Comp. 1976 a

Fritz, H., Tschesche, H., Fink, E.: Proteinase inhibitors from boar seminal plasma. In: Methods in Enzymology. S. P. Colowick and N. O. Kaplan (eds.). Vol. XLV: Proteolytic Enzymes, Part B. L. Lorand (Vol.-ed.), pp. 834–847. New York, San Francisco, London: Academic Press, 1976 b

Fuhrmann, K.: Kolorimetrische Bestimmung und histochemischer Nachweis der Aminopeptidase an Geweben weiblicher Genitalorgane. Zbl. Gynäkol. 81, 1105–1123 (1959)

Ganguly, S., Sarkar, D., Ghosh, J. J.: Sialic acid and sialidase activity in human endometrial tissue, uterine fluid and plasma under different conditions of uterine dysfunction. Acta endocr. (Kbh.) 81, 574–579 (1976)

Gardner, R. L.: An investigation of inner cell mass and trophoblast tissues following their isolation from the mouse blastocyst. J. Embryol. exp. Morph. 28, 279–312 (1972)

Glenner, G. G., Folk, J. E.: Glutamyl peptidases in rat and guinea pig kidney slices. Nature 192, 338–340 (1961)

Gothié, S.: Contribution a l'étude de la membrane pellucide de l'oeuf de Lapine à l'aide du S^{35}. J. Physiol. (Paris) 50, 293–294 (1958)

Gothié, S.: Répartition du S^{35} au cours de la nidation après injection de $^{35}SO_4Na_2$. Bull. Soc. roy. belge Gynéc. Obstét. 6, 625–634 (1960)

Gottschalk, A.: The molecular structure of ovine submaxillary gland glycoproteins. In: Salivary Glands and their Secretions. L. M. Sreebny, and J. Meyer (eds.) pp. 351–364. Oxford 1964

Gottschalk, A., Fazekas de St. Groth, S.: Studies on mucoproteins. III. The accessability to New York: Elsevier 1966

Gottschalk, A., Groth, St. de, S. Fazekas: Studies on mucoproteins. III. The accessability to trypsin of the susceptible bonds in ovine submaxillary gland mucoprotein. Biochim. Biophys. Acta (Amst.) 43, 513–519 (1960)

Gottschweski, G. H. M., Zimmermann, W.: Embryologische Untersuchungsmethoden für Laboratoriumssäugetiere. Unter Mitarbeit von H. W. Denker und U. Petzoldt. Hannover: M & H. Schaper 1970

Gottschewski, G. H. M., Zimmermann, W.: Die Embryonalentwicklung des Hauskaninchens. Normogenese und Teratogenese. Hannover: M. & H. Schaper 1973

Gräfenberg, E.: Der Antitrypsingehalt des mütterlichen Blutserums während der Schwangerschaft. Münch. med.Wschr. 56, 702–704 (1909)

Gräfenberg, E.: Beiträge zur Physiologie der Eieinbettung. Z. Geburtsh. Gyn. 65, 1–35 (1910)

Greulich, W. W.: Artificially induced ovulation in the cat (Felis domestica). Anat. Rec. 58, 217–224 (1934)

Gros, G.: Recherches préliminaires sur le cycle génital chez la Chatte. Bull. d'Histol. appl. Techn. microscop. (Paris) 10, 5–11 (1933)

Gross, F., Schaechtelin, G., Ziegler, M., Berger, M.: A renin-like substance in the placenta and uterus of the rabbit. Lancet 1, 914–916 (1964)

111

Gupta, J.S., Dey, S. K., Deb, C.: Histochemical studies on leucine aminopeptidase activity in the rat uterus. J. Reprod. Fertil. **34**, 467–473 (1973)

Hafez, E. S. E., Evans, T. N.: Human Reproduction. Conception and Contraception. Hagerstown (Maryland). New York: Harper & Row 1973

Hafez, E. S. E., Sugawara, S.: Maternal effects on some biochemical characteristics of the blastocyst in the domestic rabbit. J. Morph. **124**, 133–142 (1968)

Hagenmaier, H. E.: Zum Schlüpfprozess bei Fischen. II. Gewinnung und Charakterisierung des Schlüpfsekretes bei der Regenbogenforelle (Salmo gairdneri Rich.). Experientia (Basel) **28**, 1214–1215 (1972)

Hagenmaier, H. E.: The hatching process in fish embryos. IV. The enzymological properties of a highly purified enzyme (chorionase) from the hatching fluid of the rainbow trout, Salmo gairdneri Rich. Comp. Biochem. Physiol. **49 B**, 313–324 (1974a)

Hagenmaier, H. E.: The hatching process in fish embryos. V. Characterization of the hatching protease (chorionase) from the perivitelline fluid of the rainbow trout, Salmo gairdneri Rich., as a metalloenzyme. Wilh. Roux' Arch. **175**, 157–162 (1974b)

Halban, J., Frankl, O.: Zur Biochemie der Uterusmukosa. Gynäk. Rdschau (Wien) **4**, 471–484 (1910)

Hamana, K., Hafez, E. S. E.: Disc electrophoretic patterns of uteroglobin and serum proteins in rabbit blastocoelic fluid. J. Reprod. Fertil. **21**, 555–558 (1970)

Hamner, C. E., Jennings, L. L., Sojka, N. J.: Cat (Felis catus L.) spermotozoa require capacitation. J. Reprod. Fertil. **23**, 477–480 (1970)

Hanson, H. T., Smith, E. L.: The application of peptides containing β-alanine to the study of the specificity of various peptidases. J. Biol. Chem. **175**, 833–848 (1948)

Haour, F., Saxena, B. B.: Detection of a gonadotropin in rabbit blastocyst before implantation. Science **185**, 444–445 (1974)

Harper, M. J. K., Prostkoff, B., Reeve, R. J.: Implantation and embryonic development in the ovariectomized hamster. Acta endocr. (Kbh.) **52**, 465–470 (1966)

Hause, L. L., Pattillo, R. A., Sances jr., A., Mattingly, R. F.: Cell surface coatings and membrane potentials of malignant and nonmalignant cells. Science **169**, 601–603 (1970)

Heape, W.: The development of the mole (Talpa europea); the formation of the germinal layers, and early development of the medullary groove and notochord. Quart. J. Micr. Sci. **23**, 412–452 (1883)

Hedlund, K., Nielsson, O., Reinius, S., Åman, G.: Attachment reaction of the uterine luminal epithelium at implantation: light and electron microscopy of the hamster, guinea-pig, rabbit and mink. J. Reprod. Fertil. **29**, 131–132 (1972)

Heimburger, N., Schwick, G.: Die Fibrinagar-Elektrophorese. 1. Mitteilung: Beschreibung der Methode. Thrombos. Diathes. haemorrh. (Stuttg.) **7**, 432–443 (1962)

Heinricius, G.: Über die Embryotrophe der Raubtiere in morphologischer Hinsicht. Anat. Hefte **50**, 115–192 (1914)

Herron, M. A., Sis, R. F.: Ovum transport in the cat and the effect of estrogen administration. Am. J. Vet. Res. **35**, 1277–1279 (1974)

Hinchliffe, J. R., El-Shershaby, A. M.: Epithelial cell death in the oil-induced decidual reaction of the pseudopregnant mouse: An ultrastructural study. J. Reprod. Fertil. **45**, 463–468 (1975)

Homes, P. V., Dickson, A. D.: Estrogen-induced surface coat and enzyme changes in the implanting mouse blastocyst. J. Embryol. exp. Morph. **29**, 639–645 (1973)

Hoorn, G. van, Denker, H.-W.: Effect of the blastocyst on a uterine amino acid arylamidase in the rabbit. J. Reprod. Fertil. **45**, 359–362 (1975)

Hopsu, V. K., Ruponen, S., Talanti, S.: Leucine aminopeptidase in the placenta of the rat. Acta histochem. (Jena) **12**, 305–309 (1961)

Hruska, J. F., Law, J. H.: Cocoonases. In: Methods in Enzymology. S. P. Colowick and N. O. Kaplan (eds.) Vol. **XIX**: Proteolytic Enzymes. G. E. Perlmann and L. Lorand (Vol.-eds.), pp. 221–226. New York, London: Academic Press 1970

Hubrecht, A. A. W.: Keimblätterbildung und Placentation des Igels. Anat. Anz. **3**, 510–515 (1888)

Hubrecht, A. A. W.: Studies in mammalian embryology. I. The placentation of Erinaceus europaeus, with remarks on the physiology of the placenta. Quart. J. micr. Sci. **30**, 283–404 (1889–1890)

Hubrecht, A. A. W.: Die Säugetierontogenese in ihrer Bedeutung für die Phylogenie der Wirbeltiere. Jena: Fischer 1909

Inoue, M., Wolf, D. P.: Solubility properties of the murine zona pellucida. Biol. Reprod. 10, 512–518 (1974a)

Inoue, M. Wolf, D. P.: Comparative solubility properties of the zonae pellucidae of unfertilized and fertilized mouse ova. Biol. Reprod. 11, 558–565 (1974b)

James, N. T.: Histochemical demonstration of oxytocinase in the human placenta, Nature 210, 1276–1277 (1966)

Janssen, A.: Über die Placenta und Paraplacenta bei der Katze. Dissertation, Hannover 1933

Jaszczak, S., Hafez, E. S. E.: Endocrine control of free amino acids content in uterine fluid in pregnant rabbits. Acta endocr. (Kbh.) 70, 409–416 (1972)

Jeffrey, J. J., Gross, J.: Isolation and characterization of a mammalian collagenolytic enzyme. Fed. Proc. 26, 670 (1967)

Johnson, M. H.: Studies using antibodies to the macromolecular secretions of the early pregnant uterus. In: Immunology in Obstetrics and Gynecology. A. Centaro and N. Carretti, (eds.), pp. 123–133. Amsterdam: Excerpta Medica, 1974

Joshi, M. S., Murray, I. M.: Immunological studies of the rat uterine fluid peptidase. J. Reprod. Fertil. 37, 361–365 (1974).

Joshi, M. S., Yaron, A., Lindner, H. R.: An endopeptidase in the uterine secretion of the proestrous rat and its relation to a sperm decapitating factor. Biochem. Biophys. Res. Comm. 38, 52–57 (1970)

Kafatos, F. C.: The cocoonase zymogen cells of silk moths: A model of terminal cell differentiation for specific protein synthesis. In: Current Topics in Devel. Biol., A. A. Moscona and A. Monroy (eds.), Vol. 7, pp. 125–191. New York, London: Academic Press 1972

Kaighn, M. E.: A biochemical study of the hatching process in Fundulus heteroclitus. Develop. Biol. 9, 56–80 (1964)

Katagiri, C.: Properties of the hatching enzyme from frog embryos. J. Exp. Zool. 193, 109–118 (1975)

Kent, P.W.: Structure and function of glycoproteins. In: Essays in Biochemistry. P. N. Campbell and G. D. Greville (eds.), Vol. 3, pp. 105–151. London, New York: Academic Press 1967

Kezdy, F. J., Kaiser, E.T.: Principles of active site titration of proteolytic enzymes. In: Methods in Enzymology. S. P. Colowick and N. O. Kaplan (eds.) Vol. XIX: Proteolytic Enzymes. G. E. Perlmann and L. Lorand (eds.), pp. 3–20. New York, London: Academic Press 1970

Khera, K. S.: Teratogenic effects of methylmercury in the cat: note on the use of this species as a model for teratogenicity studies. Teratology 8, 293–304 (1973)

Kirchner, C.: Untersuchungen an uterusspezifischen Glykoproteinen während der frühen Gravididät des Kaninchens Oryctolagus cuniculus. Wilhelm Roux'Archiv Entwickl.-Mech. Org. 164, 97–133 (1969)

Kirchner, C.: Uterine protease activities and lysis of the blastocyst covering in the rabbit. J. Embryol. exp. Morph. 28, 177–183 (1972a)

Kirchner, C.: Immune histologic studies on the synthesis of a uterine-specific protein in the rabbit and its passage through the blastocyst coverings. Fertil. Steril. 23, 131–136 (1972b)

Kirchner, C.: Interferenzkontrastmikroskopische Untersuchungen über die Lyse der Keimhüllen beim Kaninchen. Cytobiol. 7, 437–441 (1973)

Kirchner, C.: Vorbereitungen auf die Implantation beim Kaninchen. Verh. Dtsch. Zool. Ges. 1974. Stuttgart: Fischer, pp. 142–145, 1975

Kirchner, C., Hirschhäuser, C., Kionke, M.: Protease activity in rabbit uterine secretion 24 hours before implantation. J. Reprod. Fertil. 27, 259–260 (1971)

Kirchner, C., Mootz, U.: Untersuchungen uber die Auflösung der Keimhüllen beim Kaninchen: Der Einfluß des uterinen Milieus. Wilh. Roux'Arch. Entwickl.-Mech. Org. 174, 172–180 (1974)

Kirchner, C., Seitz, K.-A.: Elektronenmikroskopische Untersuchungen über die Blastozyste des Kaninchens vor der Implantation in bezug auf ihre Wechselbeziehung zur uterinen Umgebung. Wilh. Roux'Arch. Entwickl.-Mech. Org. 170, 221–233 (1972)

Klein, M.: La muqueuse utérine de la Lapine. Contribution à l'histophysiologie des muqueuses. Bull. d'Histologie Appl. 10, 327–354 (1933)

Koutský, J., Rybák, M., Jirásek, J. E., Hladovec, J.: The content of plasminogen-activator in the endometrium, estimated by the fibrin-agar plate method. Gynaecologia (Basel) 167, 257–264 (1969)

113

Krüger, J.: Die Implantation des Keimes in die Uteruswand. Eine historische Betrachtung unter besonderer Berücksichtigung des Kieler Anatomen Ferdinand Graf von Spee. Diss. Med. Fakultät, Kiel 1969

Kühnel, W., Beier, H. M., Petry, G.: Untersuchungen zur hormonellen Regulation der Praeimplantationsphase der Graviditat. II. Histologische, topochemische und biochemische Analysen am hormonbehandelten, pseudograviden Kaninchenuterus. Cytobiol. 4, 9–40 (1971)

Kuhl, H., Taubert, H.-D.: Inactivation of luteinizing hormone releasing hormone by rat hypothalamic L-cystine arylamidase. Acta endocr. (Kbh.) 78, 634–648 (1975a)

Kuhl, H., Taubert, H.-D.: Short-loop feedback mechanism of luteinizing hormone: LH stimulates hypothalamic L-cystine arylamidase to inactivate LH-RH in the rat hypothalamus. Acta endocr. (Kbh.) 78, 649–663 (1975b)

Kulangara, A. C.: Volume and protein concentration of rabbit uterine fluid. J. Reprod. Fertil. 28, 419–425 (1972)

Lackie, J. M., Armstrong, P. B.: Studies on intercellular invasion in vitro using rabbit peritoneal neutrophil granulocytes. II. Adhesive interaction between cells. J. Cell. Sci. 19, 645–652 (1975)

Lammes, F. B.: Het endometrium van de muis. Diss., Leiden, 1963

Larsen, J. F.: Electron microscopy of the implantation site in the rabbit. Amer. J. Anat. 109, 319–334 (1961)

Larsen, J. F.: Electron microscopy of the uterine epithelium in the rabbit. J. Cell. Biol. 14, 49–64 (1962)

Larsen, J. F.: Histology and fine structure of the avascular and vascular yolk sac placentae and the obplacental giant cells in the rabbit. Amer. J. Anat. 112, 269–283 (1963)

Larsen, J. F.: Electron microscopy of nidation in the rabbit and observations on the human trophoblastic invasion. In: Ovo-Implantation. Human Gonadotropins and Prolactin. Second Intern. Seminar on Reprod. Physiol. and Sexual Endocrinol., Brussels, 1968. P. O. Hubinont et al. (eds.), pp. 38–51. Basel, München, New York: S. Karger 1970

Larsson, B., Liedholm, P., Åstedt, B.: Increased fibrinolytic activity in the endometrium of patients using copper-IUD (Gravigard). Int. J. Fertil. 20, 77–80 (1975a)

Larsson, B., Liedholm, P., Åstedt, B.: Effect of copper and plastic intra-uterine devices on the fibrinolytic activity of the endometrium in the rat. Int. J. Fertil. 20, 145–150 (1975b)

Leiser, R.: Kontaktaufnahme zwischen Trophoblast und Uterusepithel während der frühen Implantation beim Rind. Anat. Histol. Embryol. 4, 63–86 (1975)

Liedholm, P., Åstedt, B.: Fibrinolytic activity of the rat ovum, appearance during tubal passage and disappearance at implantation. Int. J. Fertil. 20, 24–26 (1975)

Lilien, J. E.: Toward a molecular explanation for specific cell adhesion. In: Current Topics in Develop. Biol. A. A. Moscona and A. Monroy (eds.) 4, 169–195 (1969)

Lin, Y., Means, G. E., Feeney, R. E.: The action of proteolytic enzymes on N,N-dimethyl proteins. J. Biol. Chem. 244, 789–793 (1969)

Linford, E., Iosson, J. M.: A quantitative study of some lysosomal enzymes in the bovine endometrium during early pregnancy. J. Reprod. Fertil. 44, 249–260 (1975)

Ljungkvist, I., Nilsson, O.: Ultrastructure of rat uterine luminal epithelium at functional states compatible with implantation. Z. Anat. Entw.-Gesch. 135, 101–107 (1971)

Lutwak-Mann, C.: The rabbit blastocyst and its environment: physiological and biochemical aspects. In: Biology of the Blastocyst. R. J. Blandau (ed.), pp. 243–260. Chicago, London: Univ. of Chicago Press, 1971

Lutwak-Mann, C., Hay, M. F., Adams, C. E.: The effect of ovariectomy on rabbit blastocysts. J. Endocr. 24, 185–197 (1962)

Mäkinen, K. K., Paunio, K. U.: A histochemical method for the demonstration of aminopeptidase B activity. J. Histochem. Cytochem. 20, 192–194 (1972)

Mahadevan, S., Dillard, C. J., Tappel, A. L.: Degradation of polysaccharides, mucopolysaccharides, and glycoproteins by lysosomal glycosidases. Arch. Biochem. 129, 525–533 (1969)

Manning, J. P., Meli, A., Steinetz, B. G.: Alkaline phosphatase and β-glucuronidase activity in the rat uterus during early pregnancy. J. Endocr. 35, 385–391 (1966)

Martin, B. J., Spicer, S. S., Smythe, N. M.: Cytochemical studies of the maternal surface of the syncytiotrophoblast of human early and term placenta. Anat. Rec. 178, 769–786 (1974)

Martin, F., Lambert, R., Berard, A., Martin, M.: Action de la pepsine et de la papaine sur un mucopolysaccharide sulfate d'origine gastrique. C. R. Séances Soc. Biol. Filiales 161, 1727 (1967)

114

Masquelin, H., Swaen, A.: Premières phases du développement du placenta maternel chez le lapin. Arch. Biol. **1**, 25–44 (1880)

Maurer, H. R.: Disc electrophoresis and related techniques of polyacrylamide gel electrophoresis. Second revised and expanded edition. Berlin, New York: Walter de Gruyter 1971

Maximow, A.: Die ersten Entwicklungsstadien der Kaninchenplacenta. Arch. mikroskop. Anat. (Berlin) **56**, 699–740 (1900)

McMillan, J., Folk, J. E., Glenner, G. G.: The biochemical evaluation of a histochemically defined α-L-glutamyl-β-naphthylamidase (α-GNA) activity. J. Histochem. Cytochem. **10**, 675–676 (1962)

McLaren, A.: Delayed loss of the zona pellucida from blastocysts of suckling mice. J. Reprod. Fertil. **14**, 159–162 (1967)

McLaren, A.: The fate of the zona pellucida in mice. J. Embryol. exp. Morph. **23**, 1–19 (1970)

McLaren, A., Nilsson, O.: Electron microscopy of luminal epithelium separated by beads in the pseudopregnant mouse uterus. J. Reprod. Fertil. **26**, 379–381 (1971)

McLaughlin, K. C., Hamner, C. E.: Preliminary characterization of rabbit oviduct fluid trypsin inhibitors. Biol. Reprod. **12**, 556–565 (1975)

Meyer, J. M.: Recherches sur l'ultrastructure de la muqueuse utérine de la lapine. Arch. Anat. (Strasbourg) **53**, 1–40 (1970)

Minot, C. S.: Uterus and embryo. I. Rabbit. II. Man. J. Morph. **2**, 341–462 (1889)

Mintz, B.: Control of embryo implantation and survival. In: Schering Symp. on Intrinsic and Extrinsic Factors in Early Mammalian Development, Venice 1970. Advances in the Biosciences. **6**, pp. 317–342. G. Raspé (ed.) Oxford: Pergamon Press; Braunschweig: Vieweg 1971

Mintz, B.: Implantation-initiating factor from mouse uterus. In: Biology of Mammalian Fertilization and Implantation. K. S. Moghissi and E. S. E. Hafez (eds.), pp. 343–356. Springfield, Illinois: Charles C. Thomas 1972

Mintz, B., Gearhart, J. D.: Subnormal zona pellucida change in parthenogenetic mouse embryos. Develop. Biol. **31**, 178–184 (1973)

Moore, N. W.: Progesterone requirements for the maintenance of pregnancy in the ovariectomized ewe. J. Reprod. Fertil. **43**, 386–387 (1975)

Moscona, A. A.: Cell aggregation: Properties of specific cell-ligands and their role in the formation of multicellular systems. Develop. Biol. **18**, 250–277 (1968)

Moscona, A. A. (ed.): The Cell Surface in Development. New York: John Wiley & Sons 1974

Mossman, H. W.: The rabbit placenta and the problem of placental transmission. Am. J. Anat. **37**, 433–497 (1926)

Mossman, H. W.: Comparative morphogenesis of the fetal membranes and accessory uterine structures. Carnegie Inst. Wash. Publ. 479, Contrib. to Embryol. **26**, 129–246 (1937)

Mossman, H. W., Weisfeldt, L. A.: The fetal membranes of a primitive rodent, the thirteenstriped ground squirrel. Am. J. Anat. **64**, 59–108 (1939)

Moulton, B. C.: Ovum implantation and uterine lysosomal enzyme activity. Biol. of Reprod. **10**, 543–548 (1974)

Nelson, D. M., Smith, C. H., Enders, A. C., Donohue, T. M.: The nonuniform distribution of acidic components on the human placental syncytial trophoblast surface membrane: A cytochemical and analytical study. Anat. Rec. **184**, 159–182 (1976)

Nilsson, O.: Attachment of rat and mouse blastocysts onto uterine epithelium. Int. J. Fertil. **12**, 5–13 (1967)

Noyes, R. W.: Disorders of gamete transport and implantation. In: Pathophysiology of Gestation. N. S. Assali and C. R. Brinkman III (eds.), Vol. 1, pp. 63–143. New York, London: Academic Press 1972

Oettel, M.: Beitrag zur Entwicklung von Interzeptiva. Pharmazie **30**, 42–48 (1975)

Ogawa, N., Ohi, Y.: On the chorion and the hatching enzyme of the medaka, Oryzias latipes. Zoolog. Mag. **77**, 151–156 (1968)

Oppenheimer, S. B.: Functional involvement of specific carbohydrate in teratoma cell adhesion factor. Exp. Cell Res. **92**, 122–126 (1975)

O'Rahilly, R.: Developmental stages in human embryos. Including a survey of the Carnegie Collection. Part A: Embryos of the first three weeks (stages 1 to 9). Carnegie Inst. of Washington, Publ. No. 631. Washington: Carnegie Inst. 1973

Orsini, M. W., McLaren, A.: Loss of the zona pellucida in mice, and the effect of tubal ligation and ovariectomy. J. Reprod. Fertil. **13**, 485–499 (1967)

Orsini, M. W., Psychoyos, A.: Implantation of blastocysts transferred into progesterone treated virgin hamsters previously ovariectomized. J. Reprod. Fertil. **10**, 300–301 (1965)

Otto, K., Riesenkoenig, H.: Improved purification of cathepsin B1 and cathepsin B2. Biochim. biophys. Acta **379**, 462–475 (1975)

Owers, N. O.: Comparison of the proteolytic activity of the implanting rat and guinea-pig blastocyst. Anat. Rec. **166**, 358 (1970)

Owers, N. O.: Ingestive properties of guinea pig trophoblast grown in tissue culture: A possible lysosomal mechanism. In: The Biology of the Blastocyst. R. J. Blandau (ed.), pp. 225–241. Chicago, London: The Univ. of Chicago Press, 1971

Owers, N. O., Blandau, R. J.: Enzymatic activities of implanting guinea pig embryos. Anat. Rec. **160**, 404 (1968)

Owers, N. O., Blandau, R. J.: Proteolytic activity of the rat and guinea pig blastocyst in vitro. In: The Biology of the Blastocyst. R. J. Blandau (ed.), pp. 207–223. Chicago, London: The Univ. of Chicago Press, 1971

Oya, M., Yoshino, M., Asano, M.: Human placental aminopeptidase isozymes. Experientia **30**, 985–986 (1974)

Parkening, T. A.: An ultrastructural study of implantation in the golden hamster. I. Loss of the zona pellucida and initial attachment to the uterine epithelium. J. Anat. **121**, 161–184 (1976)

Parr, E. L.: Shedding of the zona pellucida by guinea pig blastocysts: An ultrastructural study. Biol. Reprod. **8**, 531–544 (1973)

Parr, M. B., Parr, E. L.: Uterine luminal epithelium: Protrusions mediate endocytosis, not apocrine secretion, in the rat. Biol. Reprod. **11**, 220–233 (1974)

Pearse, A. G. E.: Histochemistry. Theoretical and Applied. Third edition. Vol. 1: London: J & A. Churchill Ltd., 1968. Vol. 2: Edinburgh and London: Churchill Livingstone 1972

Perlmann, G. E., Lorand, L. (eds.): Methods in Enzymology. S.P. Colowick and N. O. Kaplan (eds.). Vol. XIX: Proteolytic Enzymes. New York, London: Academic Press 1970

Perry, J. S., Heap, R. B., Amoroso, E. C.: Steroid hormone production by pig blastocysts. Nature (Lond.) **245**, 45–47 (1973)

Petry, G., Kühnel, W., Beier, H. M: Untersuchungen zur hormonellen Regulation der Praeimplantationsphase der Gravidität. I. Histologische, topochemische und biochemische Analysen am normalen Kaninchenuterus. Cytobiologie **2**, 1–32 (1970)

Petzoldt, U.: Protein patterns of the rabbit blastocyst tissues. Cytobiologie **6**, 473–475 (1972)

Petzoldt, U.: Micro-disc electrophoresis of soluble proteins in rabbit blastocysts. J. Embryol. exp. Morph. **31**, 479–487 (1974)

Petzoldt, U., Dames, W., Gottschewski, G. H. M., Neuhoff, V.: Das Proteinmuster in frühen Entwicklungsstadien des Kaninchens. Cytobiologie **5**, 272–280 (1972)

Petzoldt, U., Briel, B., Gottschewski, G. H. M., Neuhoff, V.: Free amino acids in the early cleavage stages of the rabbit egg. Develop. Biol. **31**, 38–46 (1973)

Pinsker, M. C., Sacco, A. G., Mintz, B.: Implantation-associated proteinase in mouse uterine fluid. Develop. Biol. **38**, 285–290 (1974)

Polano, O.: Über Verschwinden einer Schwangerschaft. Ein Beitrag zur Lehre von der Blasenmole. Z. Geburtsh. **59**, 453–466 (1907)

Potts, D. M.: The ultrastructure of implantation in the mouse. J. Anat. (Lond.) **103**, 77–90 (1968)

Potts, D. M., Wilson, I. B.: The preimplantation conceptus of the mouse at 90 hours post coitum. J. Anat. (Lond.) **102**, 1–11 (1967)

Psychoyos, A.: Influence of oestrogen on the loss of the zona pellucida in the rat. Nature (Lond.) **211**, 864 (1966)

Psychoyos, A.: Endocrine control of egg implantation. In: Handbook of Physiology, Sect. 7 (Endocrinology). Vol. II (Female Reproductive System), Part 2. R. O. Greep (ed.), pp. 187–215. Washington D. C.: American Physiological Society 1973

Ravetto, C.: Histochemical identification of N-acetyl-0-diacetylneuraminic acid resistant to neuraminidase. J. Histochem. Cytochem. **16**, 663 (1968)

Rehfeld, N., Peters, J. E., Giesecke, H., Haschen, R. J.: Untersuchungen über Aminosäure-arylamidasen. II. Reinigung und Charakterisierung der Aminosäure-arylamidase-Isoenzyme aus Leber und Duodenalschleimhaut des Menschen. Acta biol. med. germ. **19**, 819–830 (1967)

116

Rehfeld, N., Schultka, R.: Histochemische Untersuchungen zum Nachweis von Leucinamino-peptidase und Aminosäurearylamidase. Acta histoch. **28**, 327–334 (1967)

Reid, P. E., Culling, C. F. A., Dunn, W. L., Clay, M. G.: The use of a transesterification technique to distinguish between certain neuraminidase resistant epithelial mucins. Histochemistry **46**, 203–207 (1976)

Reinius, S.: Ultrastructure of blastocyst attachment in the mouse. Z. Zellforsch. **77**, 257–266 (1967)

Rich, R. A.: Uterine, ovarian and adrenal proteases in the normally cycling and pseudopregnant rat. Dissertation, Utah State Univ. (USA), 1965. Diss. Abstr. **26** (6), 3457 (1965)

Richardson, K. C., Jarett, L., Finke, E. H.: Embedding in epoxy resins for ultrathin sectioning in electron microscopy. Stain Technol. **35**, 313–323 (1960)

Ringler, I.: The composition of rat uterine luminal fluid. Endocrinol. **68**, 281–291 (1961)

Romeis, R.: Mikroskopische Technik. 16. Aufl. München, Wien: R. Oldenbourg 1968

Roseman, S.: The synthesis of complex carbohydrates by multiglycosyltransferase systems and their potential function in intercellular adhesion. Chem. Phys. Lipids **5**, 270–297 (1970)

Roth, S., McGuire, E. J., Roseman, S.: Evidence for cell-surface glycosyltransferases. Their potential role in cellular recognition. J. Cell Biol. **51**, 536–547 (1971)

Rumery, R. E., Blandau, R. J.: Loss of zona pellucida and prolonged gestation in delayed implantation in mice. In: The Biology of the Blastocyst. R. J. Blandau (ed.), pp. 115–129. Chicago, London: The Univ. of Chicago Press, 1971

Sapolsky, A. I., Woessner jr., J. F.: Multiple forms of cathepsin D from bovine uterus. J. biol. Chem. **247**, 2069–2076 (1972)

Sartor, P.: Evolution du "glycolemme" de l'épithélium utérine de la ratte progestante. Relation possible avec les phénomènes d'ovoimplantation. In: VII. Internat. Kongr. f. Tierische Fortpfl. u. Haustierbesamung, pp. 1871–1876, München 1972

Schiessler, H., Arnhold, M., Fritz, H.: Acid-stable proteinase inhibitors from human seminal plasma: Purification and characterization. In: Protides of the Biological Fluids. Proceedings of the XXIIIrd Colloquium, Brugge 1975. H. Peeters (ed.), pp. 163–169. Oxford etc.: Pergamon Press, 1976

Schlafke, S., Enders, A. C.: Cellular basis of interaction between trophoblast and uterus at implantation. Biol. Reprod. **12**, 41–65 (1975)

Schmidt, H., Walther, H., Voigt, K. D.: Der Einfluß von Östradiol auf den Gehalt des Rattenuterus an Nukleinsäuren, Protein und Enzymaktivitäten. Emzymol. Biol. et Clin. **7**, 239–248 (1966)

Schmidt-Matthiesen, H.: Histochemische Studien am Sekret der Endometriumdrüsen. Acta histoch. **16**, 28–45 (1963)

Schmidt-Matthiesen, H.: Die fibrinolytische Aktivität von Endometrium und Myometrium, Decidua und Plazenta, Kollum- und Korpuskarzinomen. Physiologie, Pathologie und klinisch-therapeutische Konsequenzen. Fortschr. Geburtsh. Gyn. **31**. Basel, New York: S. Karger 1967

Schmidt-Matthiesen, H.: Endometrium und Nidation beim Menschen. Z. Geburtsh. Gyn. **168**, 113–125 (1968)

Schmidt-Matthiesen, H,: The histochemistry of the human endometrium and the problem of nidation. In: Schering Symp. on Mechanisms Involved in Conception, Berlin, 1969. G. Raspé (ed.). Advances in the Biosciences 4, pp. 291–297 Oxford etc.: Pergamon Press; Braunschweig: Vieweg 1970

Schoenfeld, H.: Contribution a l'étude de la fixation de l'oeuf des mammifères dans la cavité utérine et des premiers stades de la placentation. Arch. Biol. (Paris) **19**, 701–830 (1903)

Schumacher, G. F. B.: Alpha$_1$-antitrypsin in uterine secretions. In: Intern. Res. Conf. on Proteinase Inhibitors, Munich, 1970. H. Fritz and H. Tschesche (eds.), pp. 253–256. Berlin, New York: Walter de Gruyter 1970

Schumacher, G. F. B., Zaneveld, L. J. D.: Proteinase inhibitors in human cervical mucus and their in vitro interactions with human acrosin. In: Bayer-Sympos. V: Proteinase Inhibitors. Proceed. of the 2nd Internat. Res. Conf. H. Fritz, H. Tschesche, L. J. Greene and E. Truscheit (eds.), pp.178–186. Berlin, Heidelberg, New York: Springer-Verlag, 1974

Schwick, H. G.: Chemisch-entwicklungsphysiologische Beziehungen von Uterus zu Blastocyste des Kaninchens Oryctolagus cuniculus. Roux' Arch. Entwickl.-Mech. Org. **156**, 283–343 (1965)

Seelig, H.-P., Roemheld, R.: Untersuchungen zur histochemischen Lokalisation der Leucin- und Cystinaminopeptidase (Oxytocinase) in der Placenta. Histochemie **18**, 30–39 (1969)

Seidel, F.: Die Entwicklungsfähigkeiten isolierter Furchungszellen aus dem Ei des Kaninchens. Wilhelm Roux' Arch. Entwickl.-mech. Org. **152**, 43–130 (1960)

Semm, K.: Die Bildungsstätte der Serum-Oxytocinase. Arch. Gynäk. **191**, 57–64 (1958)

Shaw jr., S. T., Cihak, R. W., Moyer, D. L.: Fibrin proteolysis in the monkey uterine cavity: Variations with and without IUD. Nature (Lond.) **228**, 1097–1099 (1970)

Shaw jr., S. T., Jimenez, J. M., Moyer, D. L., Cihak, R. W.: Relationship of endometrial plasminogen activator to fibrin proteolysis in the uterine cavity of rhesus monkeys. Am. J. Obstet. Gynecol. **115**, 983–985 (1973)

Shaw jr., S. T., Moyer, D. L., Aaronson, D. E., Underwood, J., Forino, R. V.: Intrauterine medication with epsilon aminocaproic acid. Effect on rhesus monkeys wearing intrauterine devices. Contraception **11**, 395–407 (1975)

Sherman, M. I., Salomon, D. S.: The relationships between the early mouse embryo and its environment. In: The Developmental Biology of Reproduction. 33rd Symp. Soc. for Develop. Biol. C. L. Markert and J. Papaconstantinou (eds.), pp. 277–309. New York, San Francisco, London: Academic Press 1975

Small, C. W., Watkins, W. B.: Oxytocinase: immunohistochemical demonstration in the immature and term human placenta. Cell Tiss. Res. **162**, 531–539 (1975)

Smith, A. F., Wilson, I. B.: Cell interaction at the maternal-embryonic interface during implantation in the mouse. Cell. Tiss. Res. **152**, 525–542 (1974)

Smith, E. L.: The peptidases of skeletal, heart and uterine muscle. J. Biol. Chem. **173**, 553–569 (1948)

Smithberg, M.: The effect of different proteolytic enzymes on the zona pellucida of mouse ova. Anat. Rec. **117**, 554 (1953)

Snellman, O.: Cathepsin B, the lysosomal thiol proteinase of calf liver. Biochem. J. **114**, 673–678 (1969)

Sojka, N. J., Jennings, L. L., Hamner, C. E.: Artificial insemination in the cat (Felis catus L.). Lab. Animal Care **20**, 198–204 (1970)

Somerville, M., Dabich, D.: Changes in trypsin like inhibitors in mouse uteri during early gestation. Fed. Proc. **33**, 282a (1974)

Spee, F. Graf von: Beitrag zur Entwickelungsgeschichte der frühen Stadien des Meerschweinchens bis zur Vollendung der Keimblase. Arch. Anat. Entwickl.-Gesch. **1883**, 44–60

Spee, F. Graf von: Die Implantation des Meerschweincheneies in die Uteruswand. Z. Morph. Anthropol. **3**, 130–182 (1901)

Stambaugh, R., Seitz jr., H. M., Mastroianni jr., L.: Acrosomal proteinase inhibitors in rhesus monkey (Macaca mulatta) oviduct fluid. Fertil. and Steril. **25**, 352–357 (1974)

Starck, D.: Embryologie. Ein Lehrbuch auf allgemein biologischer Grundlage. 3. Aufl. Stuttgart: Thieme 1975

Steer, H. W.: Ultrastructure of the extraembryonic region of the preimplanted rabbit blastocyst before trophoblast knob formation. J. Anat. **106**, 263–271 (1970a)

Steer, H. W.: The trophoblastic knobs of the preimplanted rabbit blastocyst: A light and electron microscopic study. J. Anat. **107**, 315–325 (1970b)

Steer, H. W.: Implantation of the rabbit blastocyst: The adhesive phase of implantation. J. Anat. **109**, 215–227 (1971a)

Steer, H. W.: Implantation of the rabbit blastocyst: The invasive phase. J. Anat. **110**, 445–462 (1971b)

Stegemann, H.: Die Primereinschlußtechnik zum Enzymnachweis bis 10^{-12} g nach Polyacrylamid-Elektrophorese, dargestellt an Phosphorylasen. Fresenius' Z. Analyt. Chem. **243**, 573–578 (1968)

Steinberg, M. S.: Does differential adhesion govern self-assembly processes in histogenesis? Equilibrium configurations and the emergence of a hierarchy among populations of embryonic cells. J. exp. Zool. **175**, 395–434 (1970)

Steven, D. H.(ed.): Comparative Placentation. Essays in Structure and Function. London, New York, San Francisco: Academic Press, 1976

Steven, V. C.: Fertility control through active immunization using placenta proteins. Karolinska Symp. on Res. Methods in Reproduct. Endocrinol., 7th Symp.: Immunological Approaches to Fertility Control, 1974. Acta Endocr. (Kbh.) Suppl. **194**, 357–375 (1975)

Sundaram, K., Connell, K. G., Passantino, T.: Implication of absence of HCG-like gonadotrophin in the blastocyst for control of corpus luteum function in pregnant rabbit. Nature (Lond.) 739–741 (1975)

118

Surani, M. A. H.: Zona pellucida denudation, blastocyst proliferation and attachment in the rat. J. Embryol. exp. Morph. 33, 343–353 (1975)

Sylvén, B.: Studies on the histochemical leucine aminopeptidase reaction. VI. The selective demonstration of cathepsin B activity by means of the naphthylamide reaction. Histochemie 15, 150–159 (1968)

Sylvén, B., Snellman. O.: Studies on the histochemical leucine aminopeptidase reaction. V. Cathepsin B as a potential effector of LNA hydrolysis. Histochemie 12, 240–243 (1968)

Sylvén, B., Snellman, O., Sträuli, P.: Immunofluorescent studies on the occurrence of cathepsin B at tumor cell surfaces. Virchows. Arch. B Cell Path. 17, 97–112 (1974)

Tachi, S., Tachi, C., Lindner, H. R.: Ultrastructural features of blastocyst attachment and trophoblastic invasion in the rat. J. Reprod. Fertil. 21, 37–56 (1970)

Tamura, Y., Niinobe, M., Arima, T., Okuda, H., Fujii, S.: Aminopeptidases and arylamidases in normal and cancer tissues in humans. Cancer Res. 35, 1030–1034 (1975)

Tappel, A. L.: Lysosomal enzymes and other components. In: Lysosomes in Biology and Pathology. J. T. Dingle and H. B. Fell (eds.) Vol. 2, pp. 207–244. Amsterdam, London: North Holland Publ. Co.; New York: John Wiley & Sons Inc., 1969

Thiery, M., Willighagen, R. G. J.: An enzymatic-histochemical study of the corpus uteri of the mouse. Anat. Rec. 146, 263–279 (1963)

Trautschold, I., Werle, E., Zickgraf-Rüdel, G.: Über dem Kallikrein-Trypsin-Inhibitor. Arzneim.-Forsch. (Drug Res.) 16, 1507–1515 (1966)

Tschesche, H., Kupfer, S., Klauser, R., Fink, E., Fritz, H.: Structure, biochemistry and comparative aspects of mammalian seminal plasma acrosin inhibitors. In: Protides of the Biological Fluids. Proceedings of the XXIIIrd Colloquium, Brugge 1975. H. Peeters (ed.), pp. 255–266. Oxford: Pergamon Press 1976

Umezawa, H.: Enzyme Inhibitors of Microbial Origin. Baltimore, London, Tokyo: University Park Press; Tokyo: University of Tokyo Press, 1972

Vogel, R., Trautschold, I., Werle, E.: Natürliche Proteinasen-Inhibitoren. (Serie: Biochemie und Klinik) Stuttgart: Thieme 1966

Vollrath, L.: Das Enzymmuster der Meerschweinchenplazenta und seine Veränderung im Verlauf der Schwangerschaft. Histochemie 4, 397–419 (1965)

Wallner, O., Fritz, H., Hochstrasser, K.: An acid-stable proteinase inhibitor in human cervical mucus. In: Protides of the Biological Fluids. Proceedings of the XXIIIrd Colloquium, Brugge 1975. H. Peeters (ed.), pp. 177–182. Oxford: Pergamon Press 1976

Walter, R., Schlank, H., Glass, J. D., Schwartz, I. L., Kerenyi, T. D.: Leucylglycinamide released from oxytocin by human uterine enzyme. Science 173, 827–829 (1971)

Wendt, V., Leidl, W., Fritz, H.: The lysis effect of bull spermatozoa on gelatin substrate film. Methodical investigations. Hoppe-Seyler's Z. Physiol. Chem. 356, 315–323 (1975a)

Wendt, V., Leidl, W., Fritz, H.: The influence of various proteinase inhibitors on the gelatinolytic effect of ejaculated and uterine boar spermatozoa. Hoppe-Seyler's Z. Physiol. Chem. 356, 1073–1078 (1975b)

Werle, E.: Zur Biochemie des Trasylol. In: Neue Aspekte der Trasylol-Therapie. G. L. Haberland und P. Matis (eds.), Vol. 3, pp. 49–59. Stuttgart, New York: Schattauer-Verlag, 1969

Werle, E.: Trasylol: Ein kurzer Überblick über Geschichte, Biochemie und Wirkungen. In: Neue Aspekte der Trasylol-Therapie. W. Brendel und G. L. Haberland (eds.), Vol. 5, pp. 9–16. Stuttgart, New York: Schattauer-Verlag, 1972

Wingender, W.: Proteinase inhibitors of microbial origin. A review. Bayer-Symp. V: Proteinase Inhibitors. Proceedings of the 2nd Internat. Research Conf. H. Fritz, H. Tschesche, L. J. Greene, and E. Truscheit (eds.), pp. 548–559. Berlin, Heidelberg, New York: Springer-Verlag, 1974

Wingender, W., von Hugo, H., Frommer, W., Schäfer, D.: A protease inhibitor isolated from Planomonospora parontospora. J. Antibiotics 28, 611–612 (1975)

Witt, I.: Biochemie der Blutgerinnung und Fibrinolyse. Weinheim: Verlag Chemie, 1975

Woessner jr., J. F.: Acid hydrolases of the rat uterus in relation to pregnancy, post partum involution and collagen breakdown. Biochem. J. 97, 855–866 (1965)

Woessner jr., J. F.: Inhibition by oestrogen of collagen breakdown in the involuting rat uterus. Biochem. J. 112, 637–645 (1969)

Woessner jr., J. F.: Purification of cathepsin D from cartilage and uterus and its action on the protein-polysaccharide complex of cartilage. J. biol. Chem. 248, 1634–1642 (1973)

Wood, J. C., Barley, V. L.: Biochemical changes in forming and regressing deciduoma in the rat uterus. J. Reprod. Fertil. **23**, 469–475 (1970)

Wood, C., Psychoyos, A.: Activité de certaines enzymes hydrolytiques dans l'endomètre et le myomètre au cours de la pseudogestation et de divers états de réceptivité utérine chez la Ratte. C. R. Acad. Sci. (Paris), Sér. D., **265**, 141–144 (1967)

Wood, J. C., Williams, E. A., Barley, V. L., Cowdell, R. H.: The activity of hydrolytic enzymes in the human endometrium during the menstrual cycle. J. Obstet. Gynaec. Brit. Cwlth. **76**, 724–728 (1969)

Yamada, K.: The reactions of sulfated polysaccharides to several histochemical tests. J. Histoch. Cytoch. **12**, 327–332 (1964)

Yamagami, K.: A method for rapid and quantitative determination of the hatching enzyme (chorionase) activity of the medaka, Oryzias latipes. Annot. Zool. Jap. **43**, 1–9 (1970)

Yamagami, K.: Isolation of a choriolytic enzyme (hatching enzyme) of the teleost, Oryzias latipes. Develop. Biol. **29**, 343–348 (1972)

Yamagami, K.: Some enzymological properties of a hatching enzyme (chorionase) isolated from the fresh-water teleost, Oryzias latipes. Comp. Biochem. Physiol. **46B**, 603–616 (1973)

Yamagami, K.: Relationship between two kinds of hatching enzymes in the hatching liquid of the medaka Oryzias latipes. J. Exp. Zool. **192**, 127–132 (1975)

Yamamoto, M.: Electron microscopy of fish development. I. Fine structure of the hatching glands of embryos of the teleost, Oryzias latipes. J. of the Faculty of Science, Univ. of Tokyo, Sec. IV, **10**, 115–121 (1963)

Yamamoto, M., Yamagami, K.: Electron microscopic studies on choriolysis by the hatching enzymes of the teleost, Oryzias latipes. Develop. Biol. **43**, 313–321 (1975)

Yasumasu, I.: Quantitative determination of hatching enzyme activity of sea urchin blastulae. J. of the Faculty of Science, Univ. Tokyo, Sec. IV, **9**, 39–47 (1960)

Yasumasu, I.: Crystallization of hatching enzymes of the sea urchin, Anthocidaris crassispina. Sci. Papers College Gen. Educ., Univ. Tokyo, **11**, 275–280 (1961)

Yasumasu, I.: Imhibition of the hatching enzyme formation during embryonic development of the sea urchin by chloramphenicol, 8-azaguanine and 5-bromo-uracil. Sci. Papers Coll. Gen. Education, Univ. Tokyo, **13**, 241–246 (1963)

Yoshizaki, N.: Ultrastructure of the frog hatching gland cell in relation to its role in the hatching process. Zool. Magazine **84**, 39–48 (1975)

Yoshizaki, N., Katagiri, C.: Cellular basis for the production and secretion of the hatching enzyme by frog embryos. J. Exp. Zool. **192**, 203–212 (1975)

Zak, B., Cohen, J.: Automatic analysis of tissue culture proteins with stable Folin reagents. Clin. Chim. Acta **6**, 665–670 (1961)

Zimmermann, W.: Methoden für experimentelle Untersuchungen am Kaninchen während der frühen Embryonalentwicklung. Zbl. f. Bakt. Paras. Infekt. und Hygiene, 1 Orig., **194**, 255–266 (1964)

Zimmermann, W.: Experimentelle Untersuchungen über die Beziehung zwischen Keim und Umwelt beim Kaninchen. Arzneimittel-Forsch. (Drug Res.) **15**, 1029–1035 (1965)

Zimmermann, W., Gottschewski, G. H. M., Flamm, H., Kunz, C.: Experimentelle Untersuchungen über die Aufnahme von Eiweiß, Viren und Bakterien während der Embryogenese des Kaninchens. Develop. Biol. **6**, 233–249 (1963)

Zsolnai, B., Somogyi, J., Szarvas, Z., Székely, L.: Die eiweißspaltenden Enzyme in der Schwangerschaft. III. Die Aktivität und subzelluläre Lokalisation von Leucinaminopeptidase, Kathepsin-B und Kathepsin-D in der unreifen und reifen Placenta. Arch. Gynäk. **205**, 1–11 (1967)

Subject Index

Other Reviews of Interest in this Series

Springer-Verlag Berlin   Heidelberg   New York